IMAGES
of America

SCHUYLKILL COUNTY FIREFIGHTING

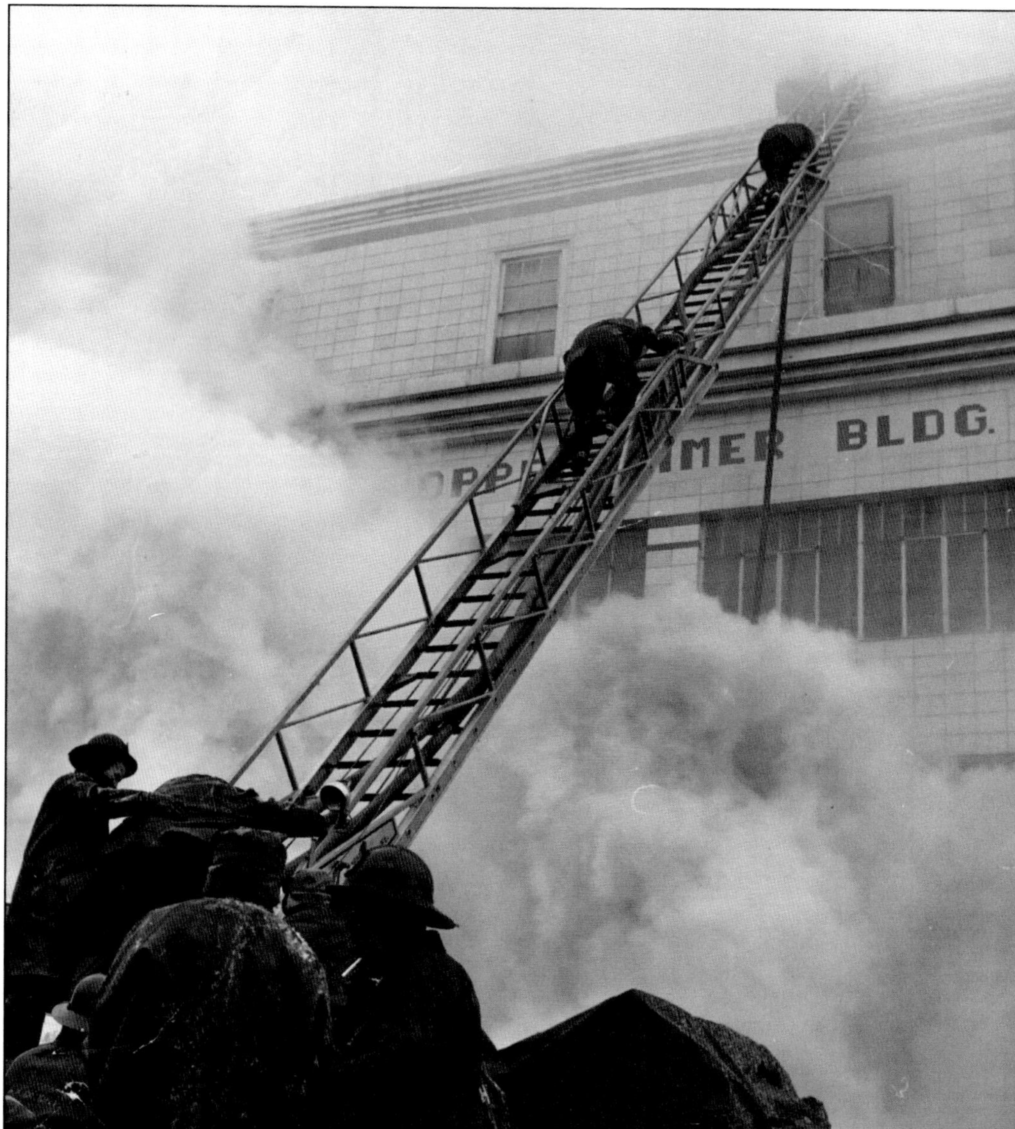

The Strand Theater was located in the four-story Oppenheimer Building at the corner of South Main and East Oak Streets in Shenandoah. On December 27, 1965, at approximately 1:32 p.m., Box 23 was transmitted for a fire that resulted from an overheated flue in a bucket-a-day coal stove in the basement of the building. In the early stages of the fire, firefighters advance hose lines over the aerial ladder on Rescue Hook and Ladder of Shenandoah's 1947 Seagrave tractor-drawn 75-foot aerial ladder truck. (Historical Society of Schuylkill County.)

On the Cover: Good Intent of Pottsville has its 1958 Oren pumper in service as West End of Pottsville operates the ladder pipe on its 1960 Seagrave 65-foot "quint" (quintuple combination pumper/ladder) at the fire that destroyed the Strand Theater on South Main Street in Shenandoah. The fire occurred December 27, 1965. Three separate fires—in 1965, 1980, and 1995—would eventually claim all the buildings in this photograph. (Historical Society of Schuylkill County.)

IMAGES
of America

SCHUYLKILL COUNTY FIREFIGHTING

Michael R. Glore and
Michael J. Kitsock

ARCADIA
PUBLISHING

ISBN 978-0-7385-7307-6

Published by Arcadia Publishing
Charleston SC, Chicago IL, Portsmouth NH, San Francisco CA

Printed in the United States of America

Library of Congress Control Number: 2009941630

For all general information contact Arcadia Publishing at:
Telephone 843-853-2070
Fax 843-853-0044
E-mail sales@arcadiapublishing.com
For customer service and orders:
Toll-Free 1-888-313-2665

Visit us on the Internet at www.arcadiapublishing.com

Contents

Acknowledgments

Volunteer firefighting commenced in Schuylkill County over 200 years ago, and the noble tradition of protecting our fellow citizens continues strongly to this day. Over 130 volunteer firefighting organizations in Schuylkill County have served their communities, and thousands of volunteers have answered the call to duty. Nowhere in the United States has the role of the volunteer firefighter been filled more proudly. This book is dedicated to generations of volunteer firefighters who faced the fiery foe and to the 35 Schuylkill County firefighters who sacrificed their lives in the line of duty.

Photograph collections from the Historical Society of Schuylkill County (HSSC) and the Schuylkill Historical Fire Society (SHFS) provided most of the materials for *Schuylkill County Firefighting*. Several Schuylkill County fire companies have also responded to a call for information. These include the Good Intent Fire Company of Llewellyn; American Fire Company, Fountain Springs; Good Will Fire Company No. 4, Pottsville; and the Phoenix Fire Company No. 2, Shenandoah.

Many brother and sister firefighters provided encouragement, support, and materials. We would be remiss not to mention some of our greatest contributors. Fire chiefs include Frank Zangari, Joseph Kufrovich, David Hardinger, Jason Kane, Eric Eichenburg, James Reed, and Michael Pothering. Firefighters and friends are Francis Senglar, Jon Morgalis, David Pothering, Thomas Nork, Thomas Dempsey, Herbert Spangler, John Yaag, James McGlaughlin, William Reigle, Raymond Jones, Adam Bernodin, and Robert Dusel. Special thanks go to Dean Meyers for the information on the June 2, 1959, Red Church disaster in Orwigsburg and also to Richard Nagle of the Historical Society of Schuylkill Haven. Schuylkill County Fire photographer Daniel Markiewicz added several fire photographs from his archives, which greatly augmented our efforts.

Enjoy your reading as we commemorate our proud heritage.

—Michael R. Glore and Michael J. Kitsock

INTRODUCTION

No area the size of Schuylkill County in the United States can boast of more volunteer firefighters serving its communities. Following the discovery of anthracite coal in Schuylkill County, Pennsylvania, in 1790, wave after wave of European immigrants flocked to the coalfields in pursuit of jobs and dreams. To protect developing neighborhoods and communities, volunteer fire companies soon emerged, each proclaiming lofty ideals through a proud name—Liberty, Friendship, Good Intent, Humane, Alert, Defenders, Independence, German-American, and Polish-American to name a few. By the mid-20th century, over 130 volunteer fire companies had been organized in Schuylkill County, serving as focal points for safety and socialization of the citizens.

Schuylkill County Firefighting presents the proud story of volunteer firefighting in this unique region of Pennsylvania. From the inception of volunteer firefighting in Schuylkill County in 1820 through the technological advancements of today, a story steeped in dedication, tradition, loyalty, bravery, and unflinching service to fellow man emerges. Firefighting is inherently dangerous, and 35 Schuylkill County volunteer firefighters have made the ultimate sacrifice with their lives. Several of these dramatic stories become apparent through the images in this book.

With over 10,000 volunteer firefighters in the county membership ranks today, *Schuylkill County Firefighting* will have magnetic appeal throughout this region. Generations of family members have dutifully served respective community volunteer fire companies. Veteran firefighting coauthors Michael Glore and Michael J. Kitsock have carefully incorporated images from every region in Schuylkill County, Pennsylvania—from the sleepy farming community of Orwin to the industrial city of Pottsville, from the mining community of Shenandoah to the old canal towns of Port Clinton and Port Carbon. As firefighters themselves, Glore and Kitsock provide firsthand accounts of many of the incidents included in this book.

Several major disasters have occurred in Schuylkill County and still hold immense local interest. These catastrophes include the 1959 Red Church propane truck explosion in West Brunswick Township, one of the nation's deadliest hazardous-material incidents, and the Memorial Day fire of 1945 in Mahanoy City, one of Pennsylvania's largest conflagrations. The Red Church propane truck explosion killed 11 people, including two firefighters, and injured many more. This past year marked the 50th anniversary of that disaster, and media interest has been extensive. The Memorial Day inferno of 1945 in Mahanoy City, caused by an arsonist, destroyed 66 buildings in the downtown business district and remains one of Pennsylvania's worst fires. Many of the images of these catastrophes have never been previously published.

The year 2011 marks the 200th anniversary of Schuylkill County, and many traditional events promise to underscore the county's firefighting traditions as part of the bicentennial celebrations. The Schuylkill County Volunteer Firefighters' Association Convention Parade, one of the largest firefighter parades in the nation, annually attracts up to 20,000 spectators. The Schuylkill Historical Fire Society Annual Antique Muster draws hundreds of firefighters from throughout Pennsylvania and beyond. The Historical Society of Schuylkill County sponsors historical and cultural events throughout the year. These events promise to provide an exciting background to the images and stories presented in *Schuylkill County Firefighting*.

One

North

With the discovery of rich anthracite deposits, northern Schuylkill County experienced a gold rush in the mid-19th to early 20th centuries. Thousands of European immigrants arrived and provided the laborers for the mining industry. Schuylkill County's coal towns developed quickly, and fire companies were soon established. The Humane Fire Company of Mahanoy City was the first to organize in 1867. Over the next 60 years, four more Mahanoy City companies formed: the Citizens in 1870, Washington Hook and Ladder in 1886, the German-American (Good American) in 1893, and the West End in 1925.

Ashland's first fire company, the Washington, started in 1867, and its second, the American, organized in 1878. Shenandoah, boasting a mining population of over 30,000 in the early 20th century, formed five fire companies: the Columbia (1870), the Rescue Hook and Ladder (1873), Phoenix (1885), Defender (1895), and the Polish-American (1915). The nearby boroughs of Girardville and Gilberton established fire departments as well. The Girard Hose Company organized in 1882; the Rangers in 1911, and the West End in 1917. Gilberton Borough created the Friendship, American, and Continental in Maizeville, Mahanoy Plane, and Gilberton in 1886.

Situated on top of Broad Mountain, Frackville launched the Good Will in 1910, and Gordon assembled a fire company in 1927. West Mahanoy Township put together the William Penn Fire Company in 1907. Other West Mahanoy Township companies included Lost Creek's West End and Citizens in 1919 and 1922, Heights Fire Company of Shenandoah Heights in 1923, and the Altamont Fire Company in 1931. Butler Township formed four fire companies, including the American of Fountain Springs in 1924, the Friendship of Englewood in 1925, Locust Dale in 1935, and Lavelle in 1937.

The farming areas of Ringtown and Union Township organized fire companies in 1898 and 1936. These two companies merged to form the Ringtown Valley Fire and Rescue Company in 2008. New Castle Township organized a fire company in Wadesville in 1920. East Union Township created the Sheppton-Oneida Fire Company in 1933 and the Citizens Fire Company of Brandonville in 1936. North Union Township formed the Nuremburg/Weston Fire Company in 1934. Kline Township initiated a fire company in 1946, which later disbanded. McAdoo activated two fire companies: the Keystone Fire Company in 1896 and the McAdoo Fire Company in 1921.

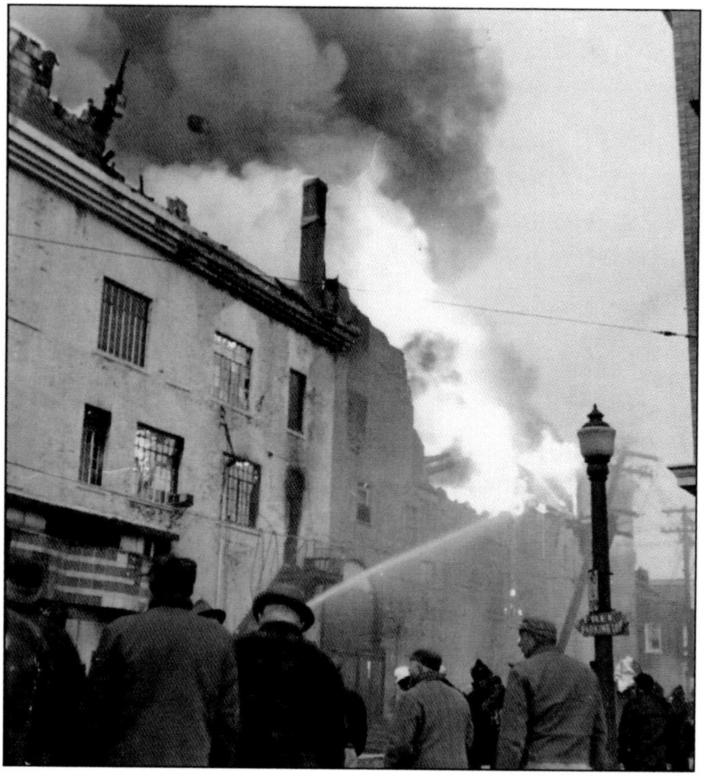

The wall along the East Oak Street side of the Oppenheimer Building begins to collapse during the fire of December 27, 1965, in Shenandoah. A tragedy was averted as nearly 300 individuals—mostly children—were evacuated from the Strand Theater, which also occupied the building. Fire in the basement broke out shortly after the 1:00 p.m. matinee had begun. (HSSC.)

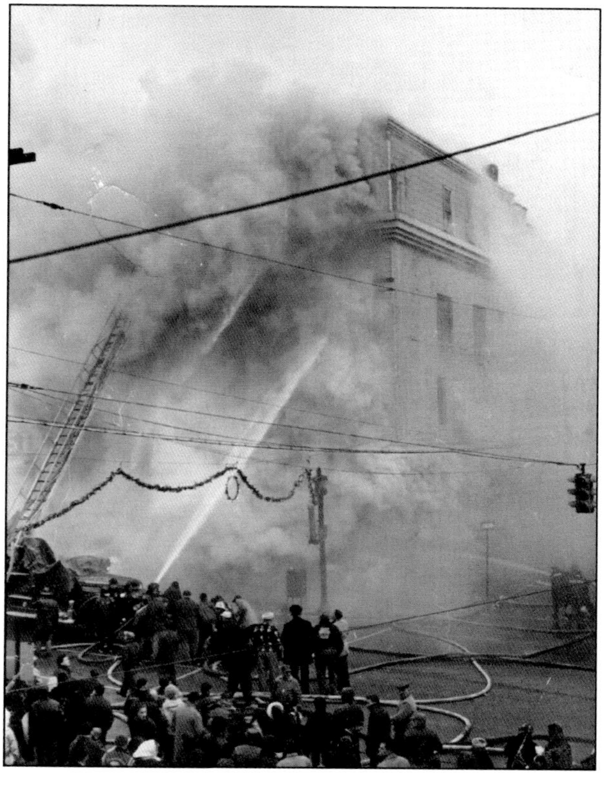

Firefighters wage a desperate battle at the corner of South Main and East Oak Streets in Shenandoah as conditions deteriorate within the Oppenheimer Building. A protective cover lies over the open cab on Rescue Hook and Ladder of Shenandoah's Seagrave aerial ladder truck. Holiday decorations stretch between the streetlamps. (HSSC.)

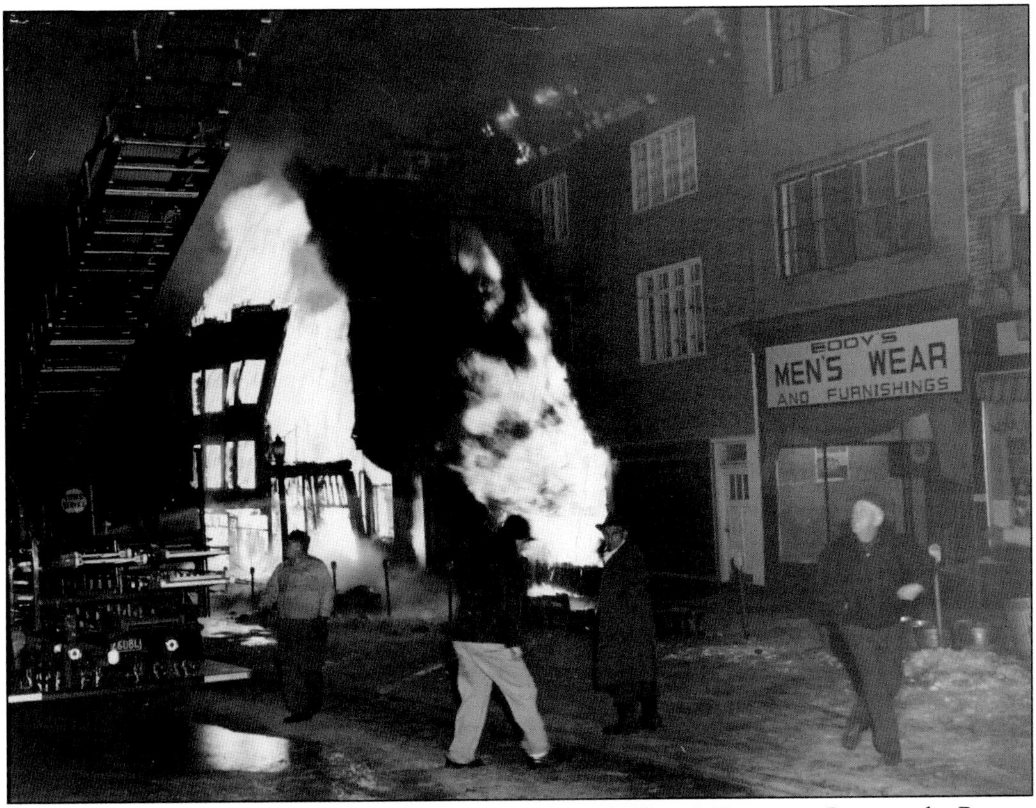

Heavy fire billows from the first-floor display window of Ufberg's Furniture Store as the Rescue Hook and Ladder Company of Shenandoah places its Seagrave tractor-drawn aerial ladder in service on South Main Street on February 4, 1955. The alarm was sounded at 2:50 a.m. All the buildings in the photograph were eventually consumed. (HSSC.)

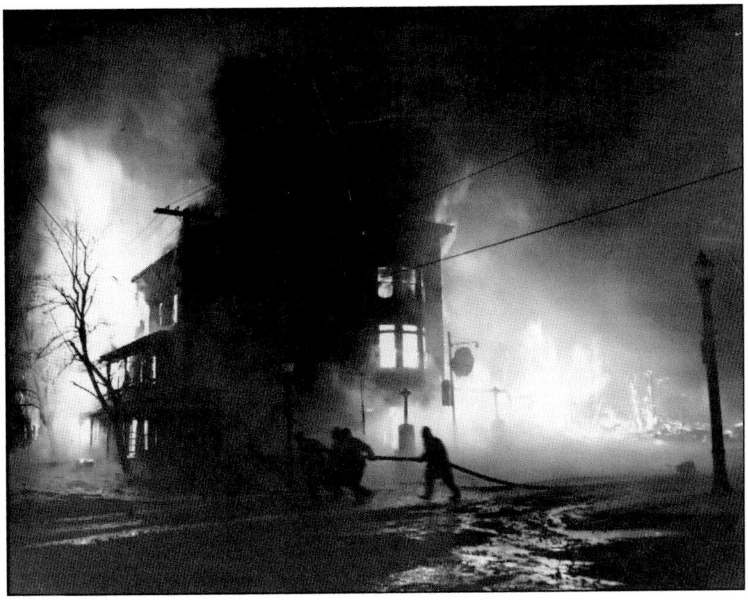

An eerie scene is depicted as firefighters work to position a 2.5-inch hose line as fire consumes the 100 block of South Main Street in Shenandoah on February 4, 1955. In addition to heavy fire conditions in the frame buildings in the block, firefighters also had to contend with temperatures hovering near 0 degrees. Fire companies from throughout northern Schuylkill County assisted at the fire. (HSSC.)

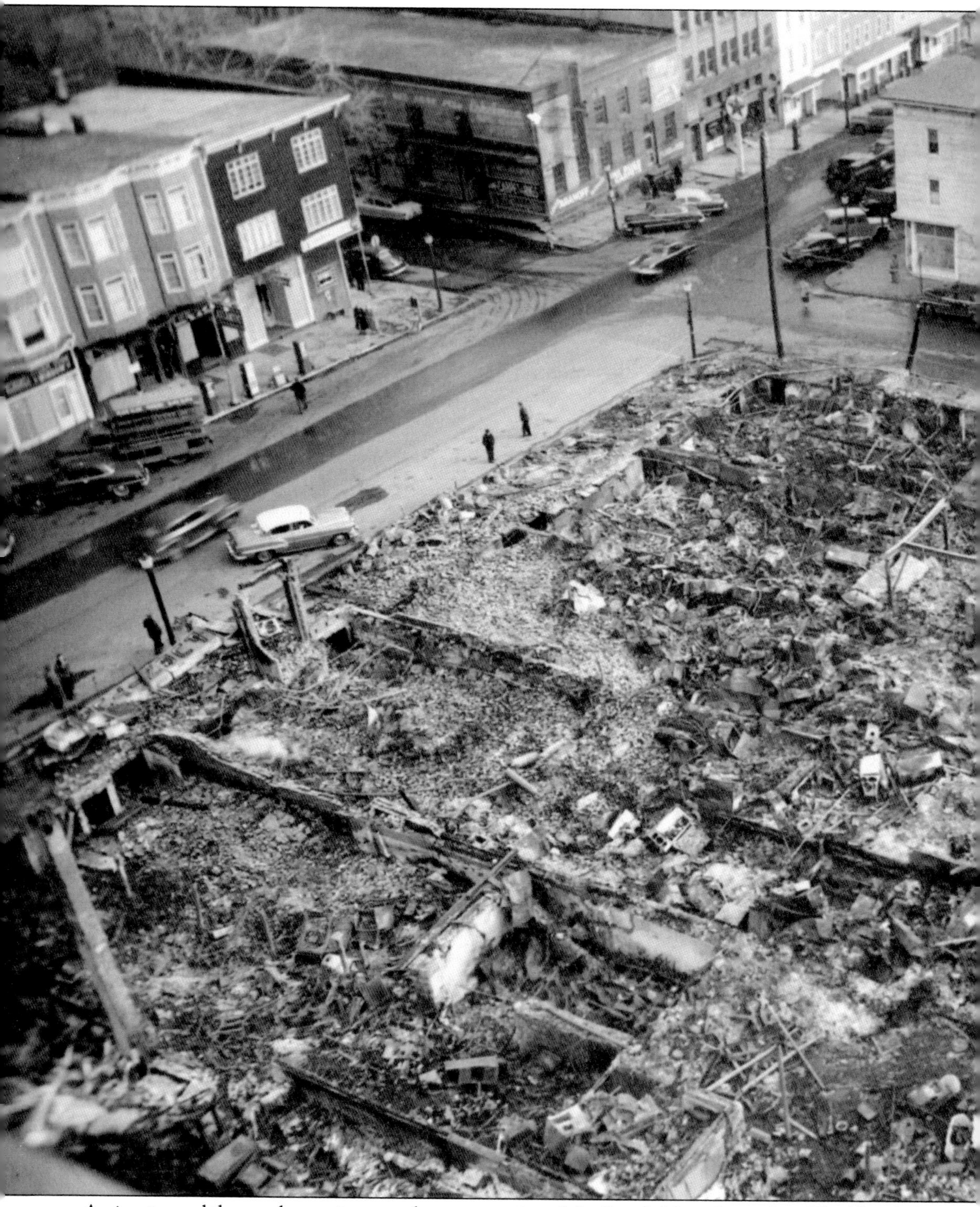

A view toward the southeast gives another perspective of the South Main Street fire. The blaze left more than 100 individuals homeless, as 44 families lost their homes. Twelve businesses were also destroyed. At the time, it was the largest fire in Shenandoah since November 12, 1883. Directly

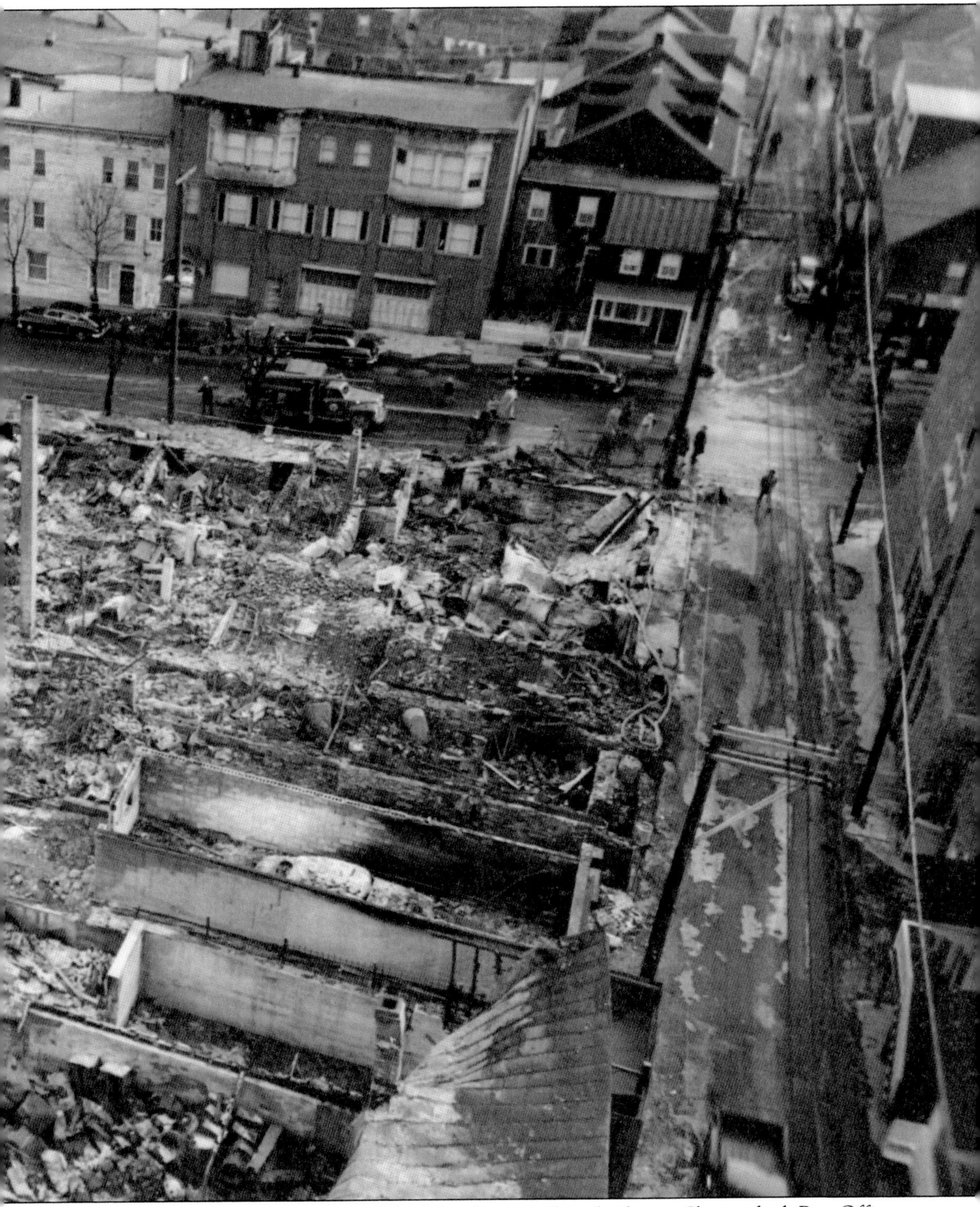

behind the Texaco sign in the upper left of the photograph is the former Shenandoah Post Office. This building was destroyed in a very spectacular fire in June 2007. (HSSC.)

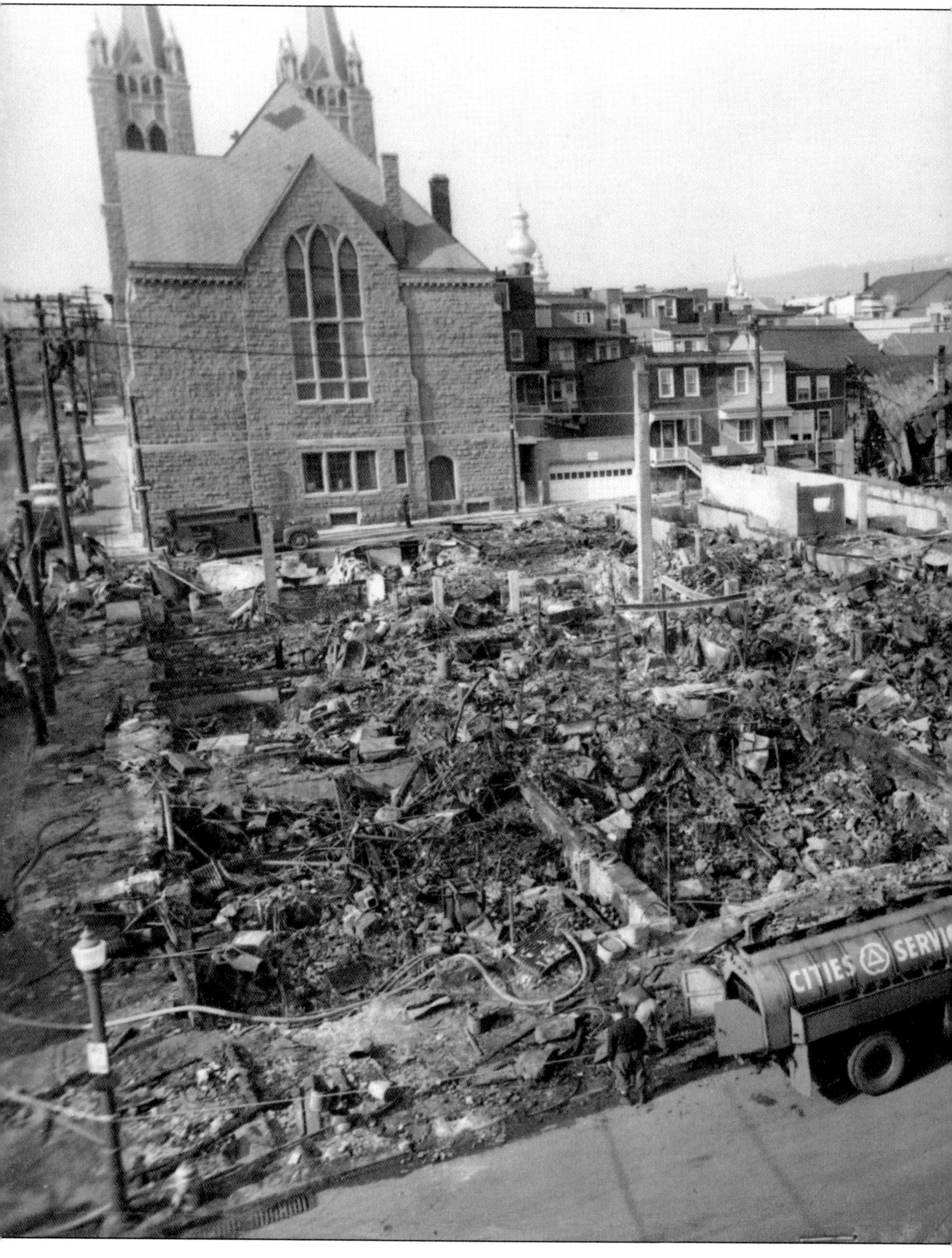
CITIES SERVIC

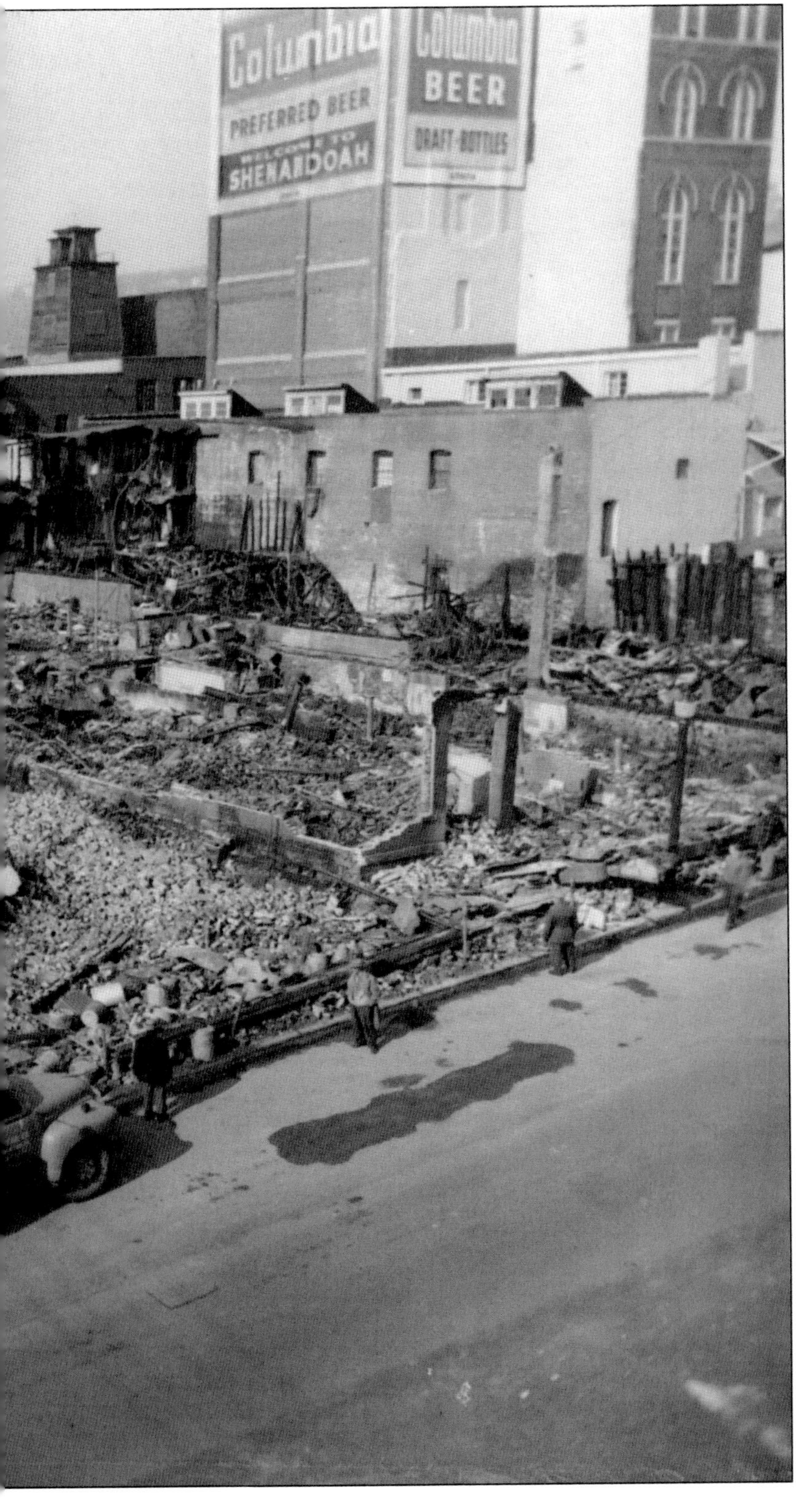

This view looking northwest clearly shows the devastation following the South Main Street fire of February 1955. Firefighters prevented the fire from extending to the church and other homes and businesses bordering the block in which the fire occurred. It is believed that the fire originated in the north wall of the Aranoff's Dry Goods store at 132 South Main Street. The building with the wooden tower on the roof just below the Columbia Beer sign is the current home of the Schuylkill Historical Fire Society and the past home of the Columbia Hose Company of Shenandoah. (HSSC.)

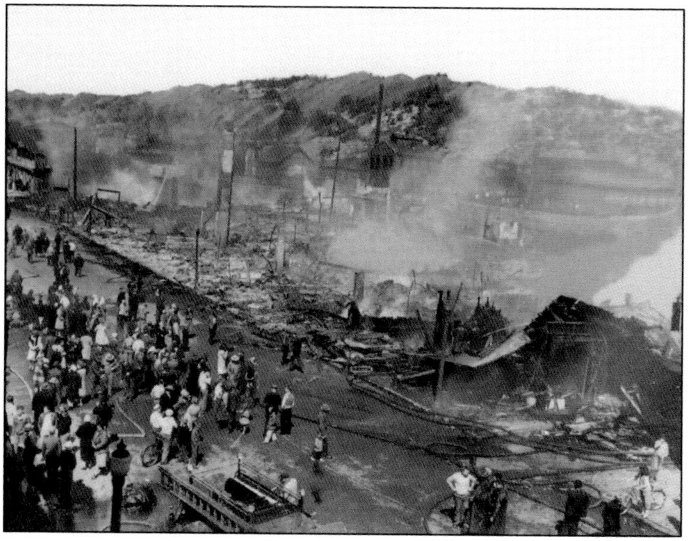

The unbelievable destruction is evident along West Center Street in Mahanoy City following a $1.5 million fire on Memorial Day, 1945. Around 2:00 a.m., an arson fire originating in a vacant garage at 126 West Railroad Street spread rapidly and destroyed nearly two square blocks. In all, 200 individuals were left homeless and 25 businesses were destroyed. Twenty-three pieces of fire apparatus from nine communities battled the fire. (Francis Senglar.)

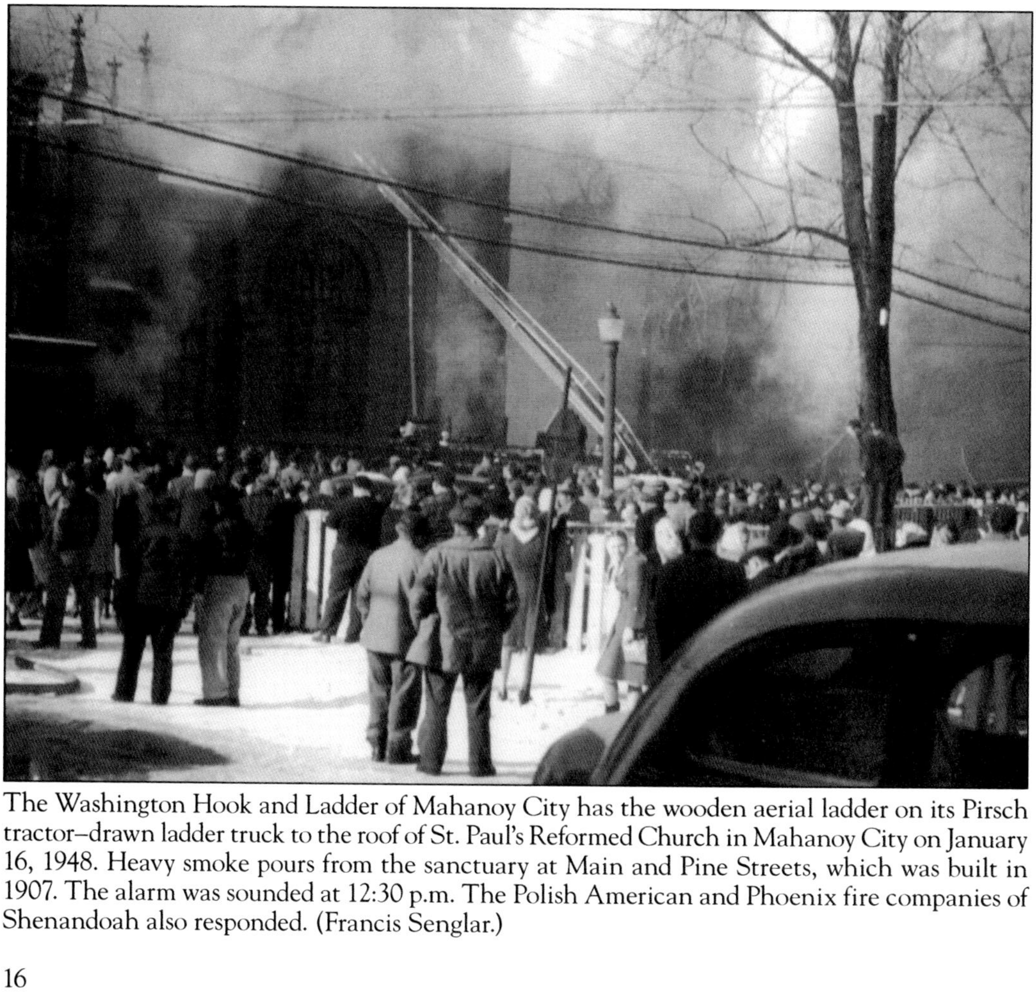

The Washington Hook and Ladder of Mahanoy City has the wooden aerial ladder on its Pirsch tractor–drawn ladder truck to the roof of St. Paul's Reformed Church in Mahanoy City on January 16, 1948. Heavy smoke pours from the sanctuary at Main and Pine Streets, which was built in 1907. The alarm was sounded at 12:30 p.m. The Polish American and Phoenix fire companies of Shenandoah also responded. (Francis Senglar.)

Rescue Hook and Ladder of Shenandoah's vintage Seagrave 75-foot tractor-drawn aerial ladder awaits a water supply to the ladder pipe at this massive fire on South Main Street in Shenandoah on January 8, 1980. Rescue Hook and Ladder's American LaFrance TeleSQURT is also in the photograph, with an elevated master stream in service. (SHFS.)

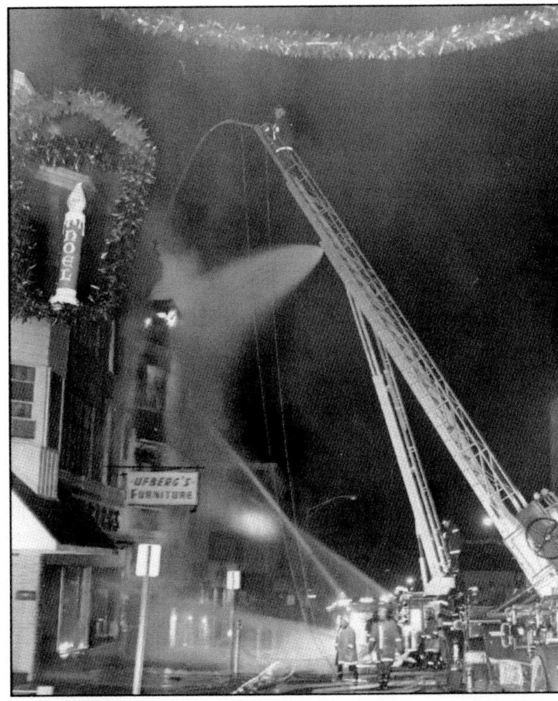

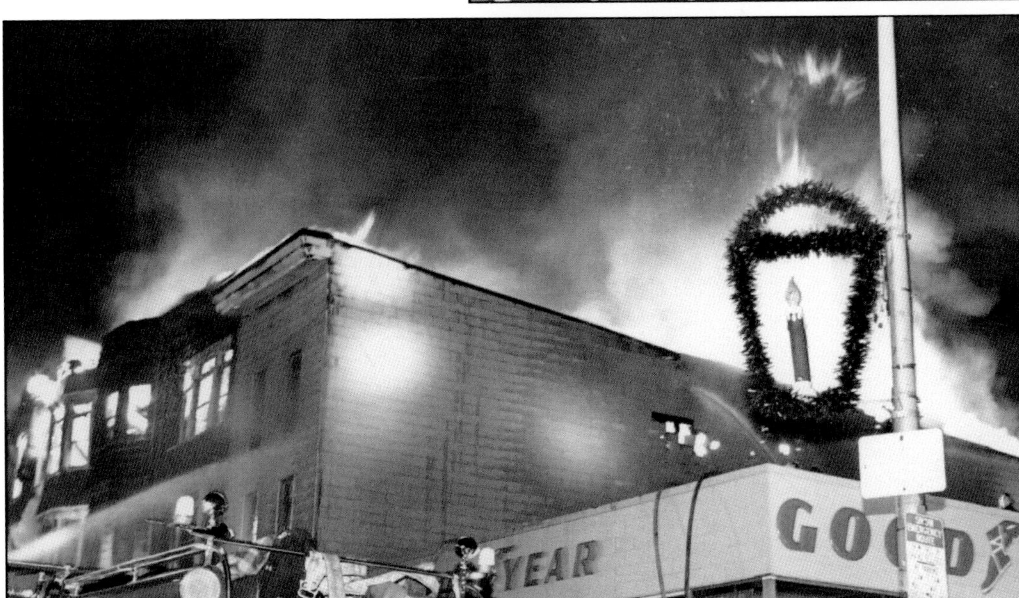

Defender Hose Company No. 3 of Shenandoah has a 1974 Hahn pumper in service at the January 8, 1980, fire on South Main Street. The three-story frame buildings are fully involved in fire. Flames silhouette a Christmas decoration on the utility pole to the right. (SHFS.)

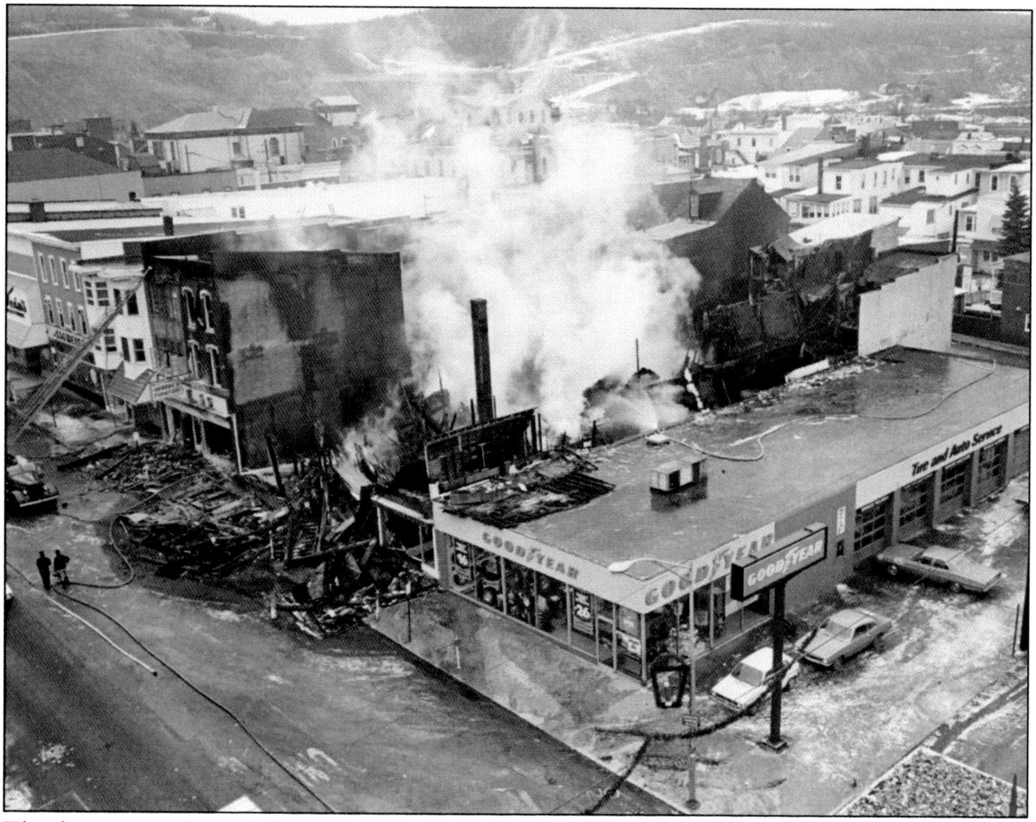

The devastation from the January 8, 1980, fire in Shenandoah is evident. This block was hit hard by fire over the years. The Goodyear tire store was the site of the Oppenheimer Building fire on December 27, 1965. The white buildings to the left behind the aerial ladder were destroyed on October 22, 1995. In each of the fires, firefighters from throughout Schuylkill County provided aid. (HSSC.)

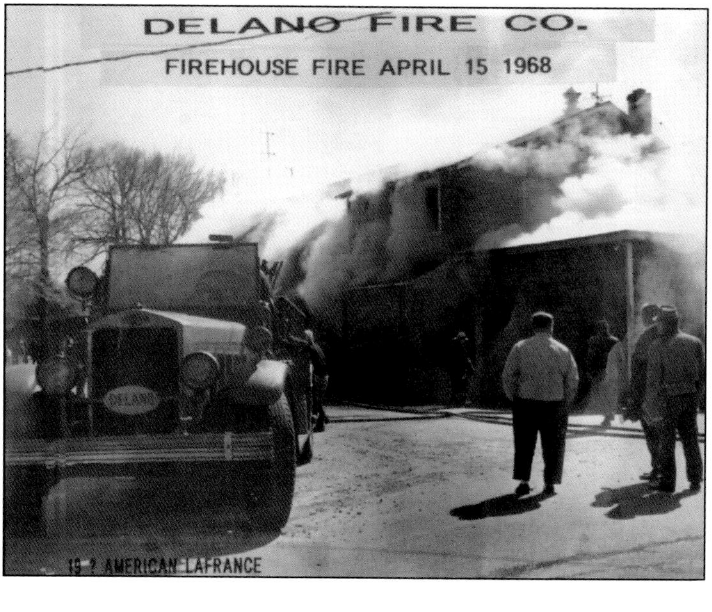

At 8:30 a.m. on April 15, 1968, fire struck the Delano Fire Company. Smoke was noticed coming from the building by an attendant in the Delano Fire Tower manned by the Pennsylvania Department of Forest and Waters. The damaged American LaFrance pumper pictured here was pulled out of the engine room using tow chains. Fire damaged totaled $70,000. (Francis Senglar.)

Rescue Hook and Ladder of Minersville has its snorkel in service on February 8, 1975, in Mahanoy City. Fire companies from throughout northern Schuylkill County, plus companies from Pottsville and Minersville, assisted the Mahanoy City Fire Department. Nineteen individuals were left homeless in the fire. (HSSC.)

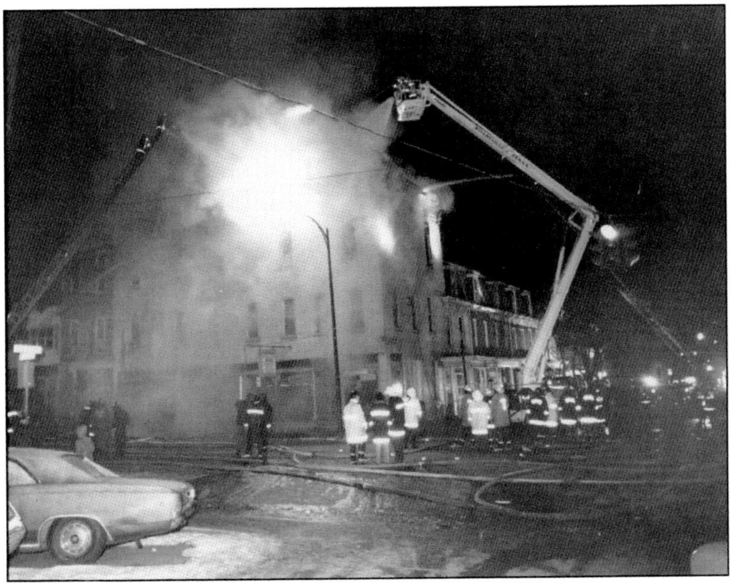

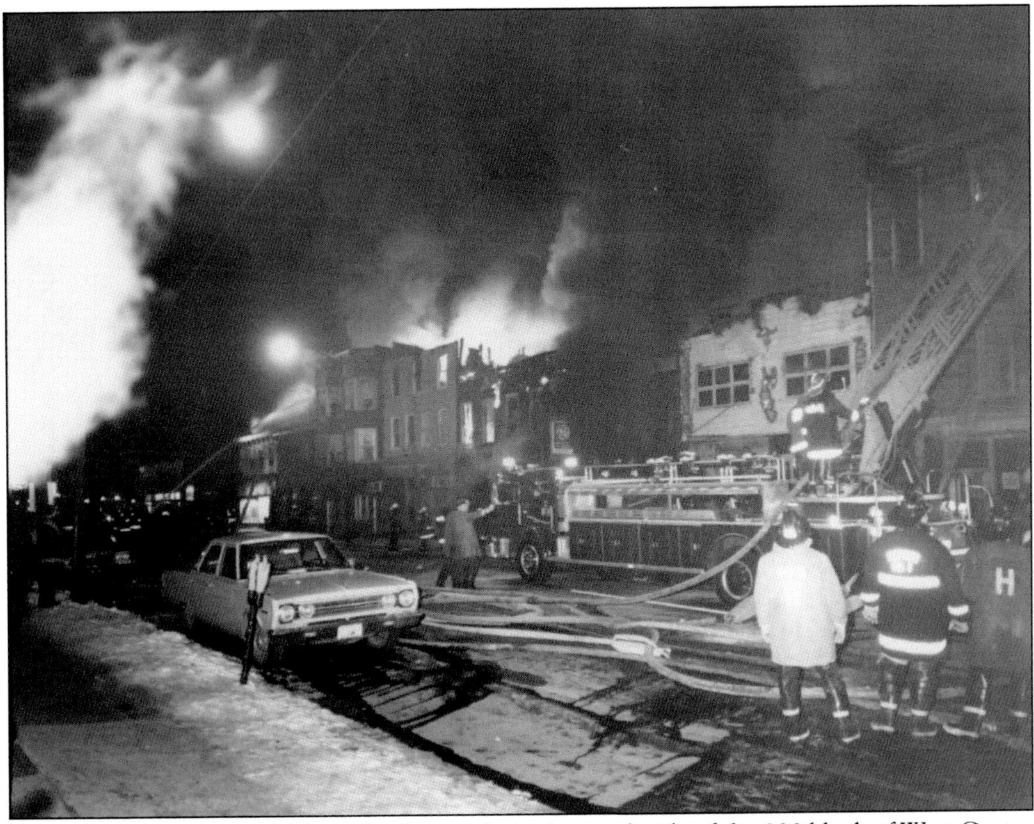

Fire consumes the block of frame buildings along the south side of the 200 block of West Centre Street in Mahanoy City. Washington Hook and Ladder of Mahanoy City has its 100-foot American LaFrance aerial ladder truck in service. The alarm sounded at 12:45 a.m. on February 8, 1975. The Citizens Fire Company's quarters now occupies much of this site. (HSSC.)

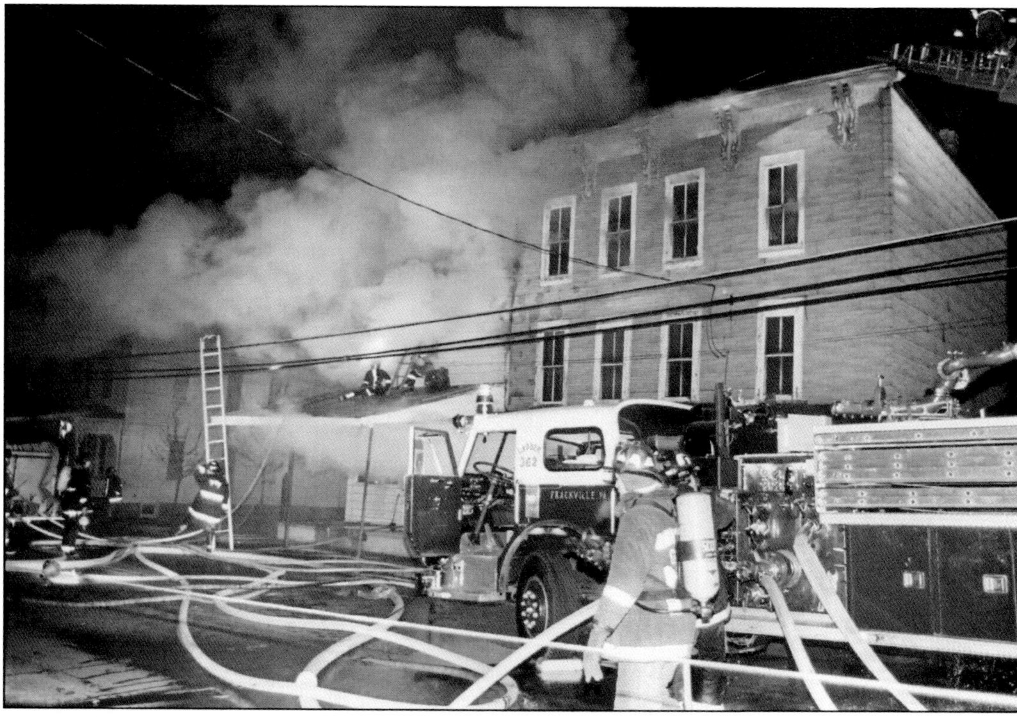

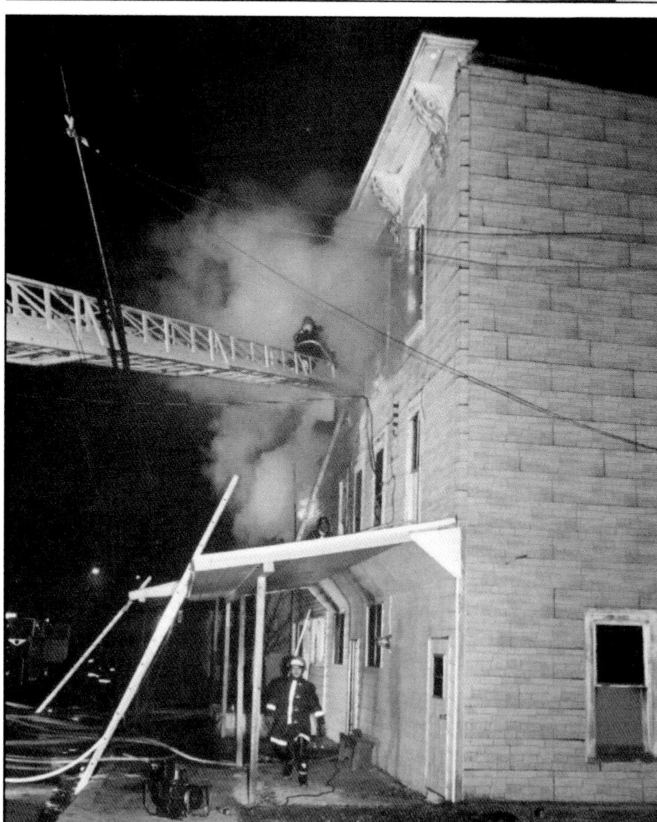

Good Will Fire Company of Frackville is in service with its 1974 American LaFrance 100-foot quint at this fire at 13–15 Main Street in Ringtown on December 2, 1983. The fire began in a vacant home and spread to the bar next door. The fire was discovered at 12:30 a.m. when bar patrons smelled smoke. (HSSC.)

Heavy smoke begins to push from the top floor after firefighters working from the tip of Frackville's aerial ladder begin to vent the building by removing the window glass. Hose lines can be seen advanced into the buildings involved in the fire on Main Street in Ringtown. (HSSC.)

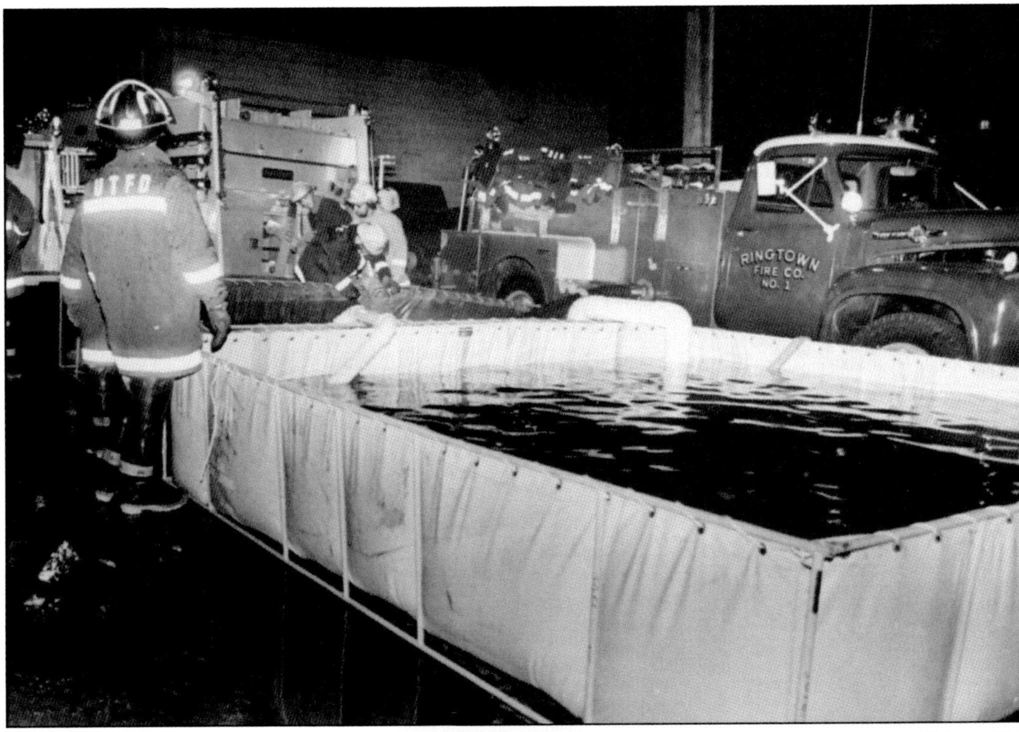

In order to establish an adequate water supply to fight the fire in Ringtown, firefighters utilized portable dump tanks supplied by tankers (now referred to as water tenders). Here Ringtown's Seagrave pumper with a commercial chassis drafts water from a portable holding tank supplied by tankers from Union Township and Sheppton. (HSSC.)

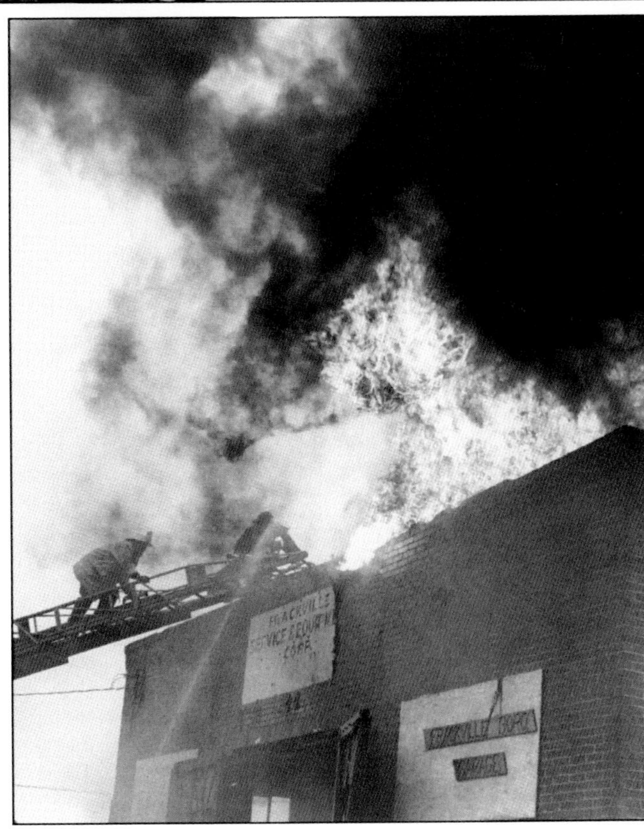

The Good Will Fire Company of Frackville had placed a new 1974 American LaFrance 100-foot aerial ladder in service shortly before fire erupted at the Frackville Borough garage on May 12, 1975. Soon after firefighters opened a ventilation hole in the roof, fire erupted from it. Firemen retreat down the ladder as flames consume the roof. (HSSC.)

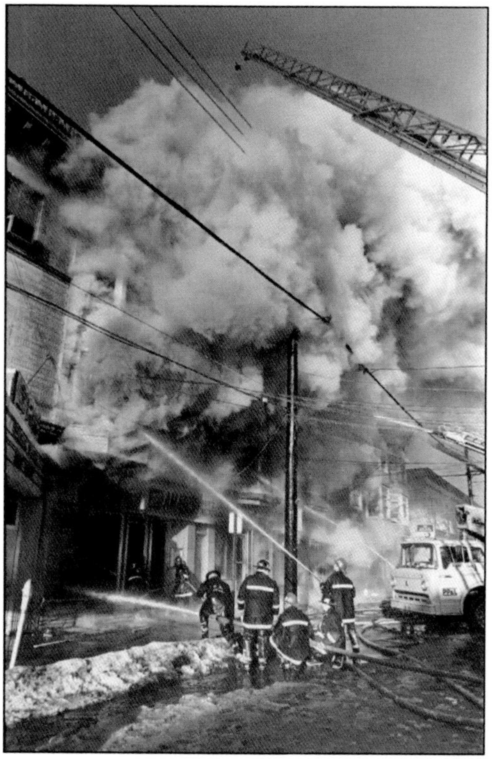

Firefighters are confronted with an ugly smoke condition at this fire, which consumed several three- and four-story frame buildings on West Centre Street in Shenandoah on February 10, 1982. Hose line streams are directed into the buildings as the Rescue Hook and Ladder of Shenandoah has a ladder pipe at the ready on its Seagrave aerial ladder. The Lyric Theater is in the extreme left of the photograph. (HSSC.)

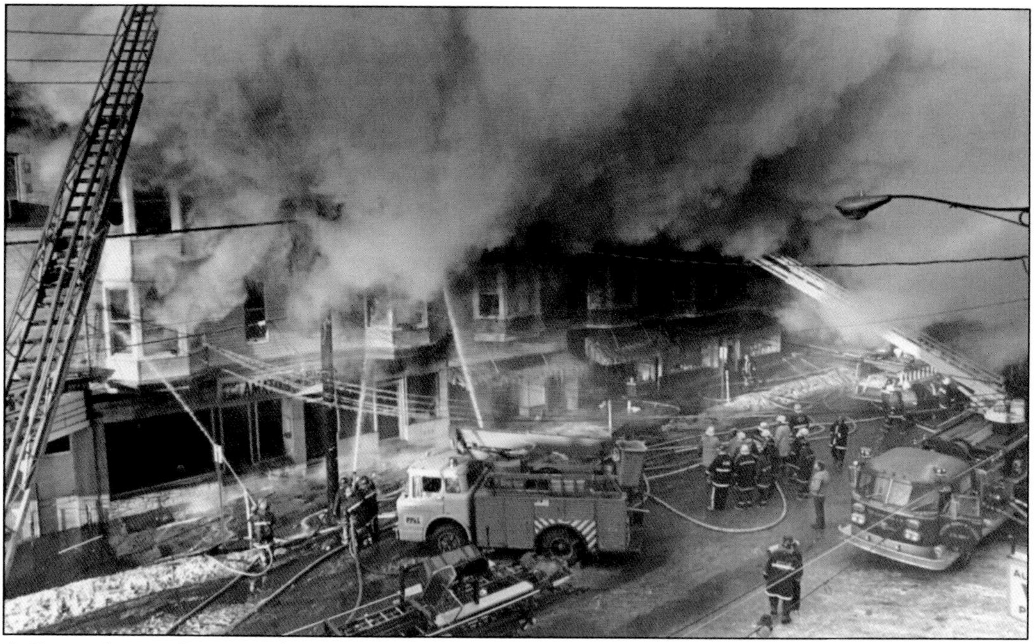

A Pennsylvania Power and Light utility truck secures electric service to the buildings involved in the February 10, 1982, Shenandoah fire. The blaze originated in Amato's Pizza shop. In addition to a Seagrave aerial ladder, Rescue Hook and Ladder of Shenandoah also has an American LaFrance TeleSQURT in service in the right of the photograph. Heavy smoke envelops the frame buildings as the fire extends. (HSSC.)

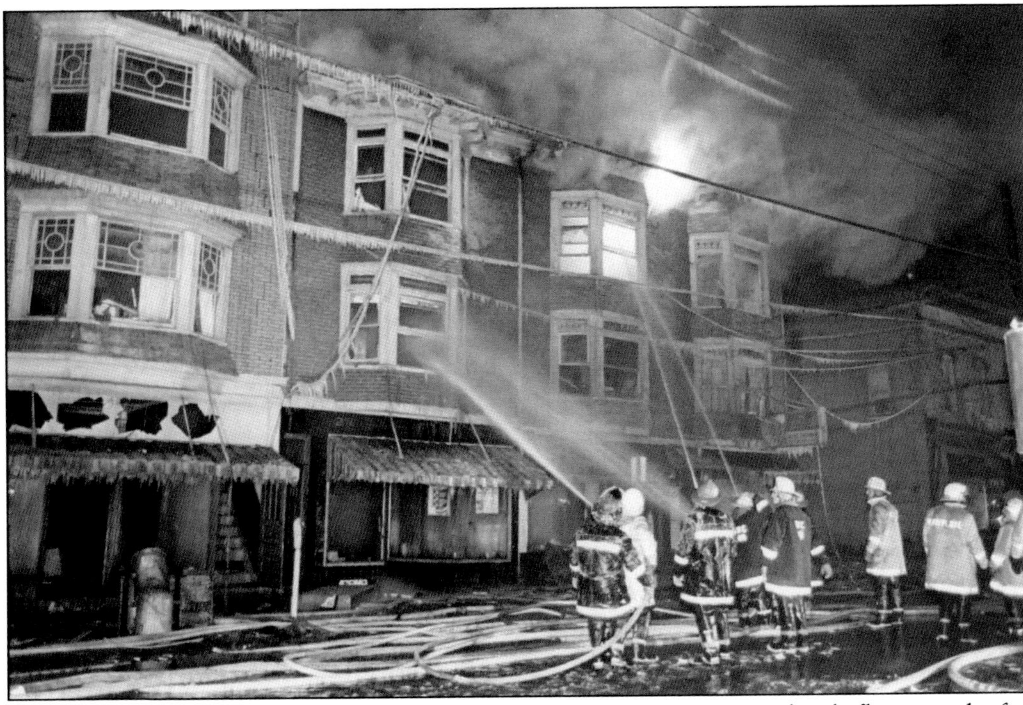

Day turns into night and the temperature falls as firefighters continue to battle flames at the fire on West Centre Street in Shenandoah on February 10, 1982. Ice is accumulating on the utility wires and on the firefighters' protective clothing. (HSSC.)

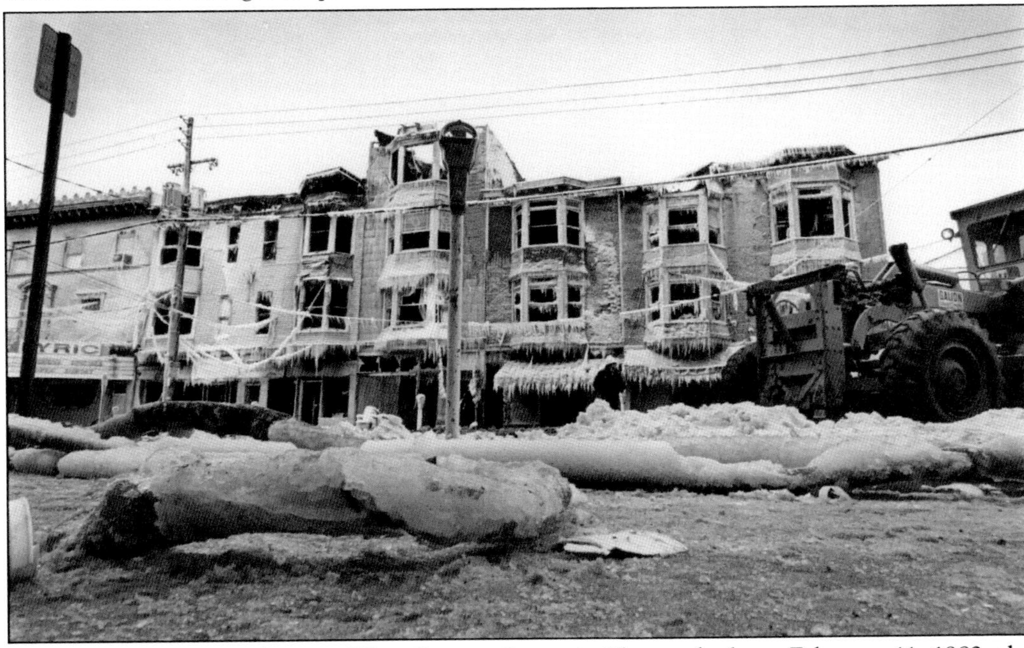

The devastation is evident on West Centre Street in Shenandoah on February 11, 1982, the day after the fire. Heavy equipment from the Pennsylvania Department of Transportation is being used to clear ice accumulation from the street. Hose lines lie frozen in the street in the foreground. (HSSC.)

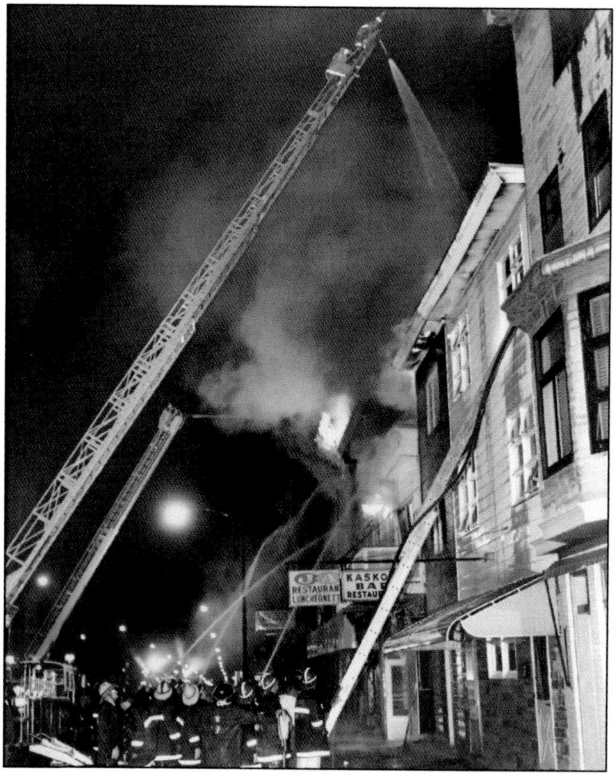

Fire extends from Davidson's Furniture Store to other buildings in the heart of Mahanoy City's business district on September 1, 1975. Elevated master streams are provided by Mahanoy City and Shenandoah aerial devices. This was the third multi-building fire that fire chief John Wertz and the Mahanoy City Fire Department faced in 1975. (HSSC.)

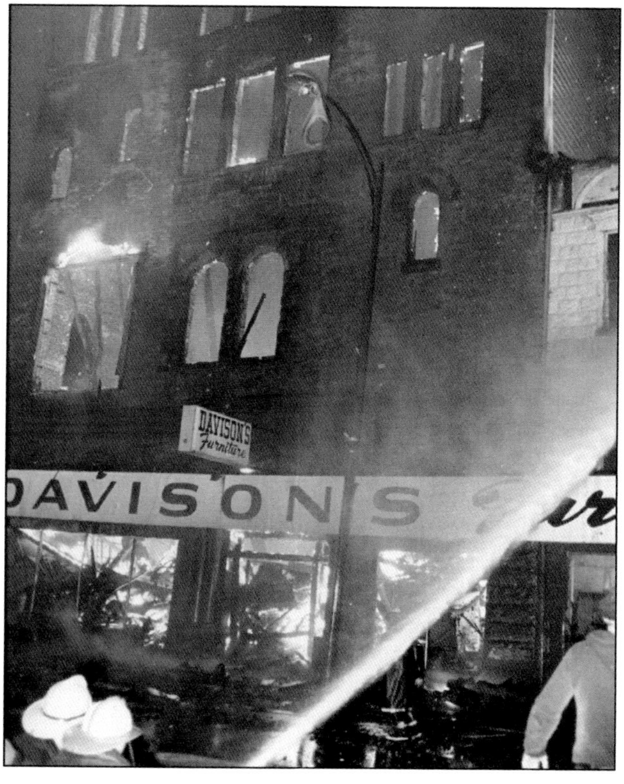

Engulfed in flames, the five-story Davidson's Furniture Store in Mahanoy City would be destroyed by this blaze on September 1, 1975. The alarm was transmitted at 11:15 p.m. Fire companies from Frackville, Ryan Township, Quakake, Delano, Shenandoah, Gilberton, Sheppton, and Tamaqua assisted the Mahanoy City Fire Department. Several other buildings were also destroyed or damaged in the fire on East Centre Street. (HSSC.)

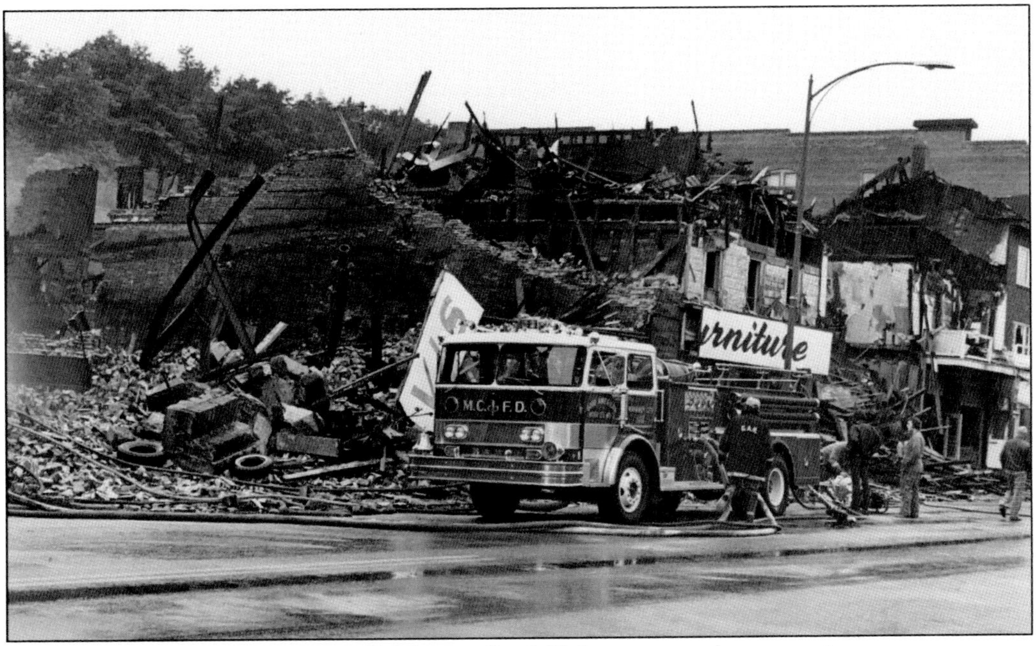

Good American Hose Company of Mahanoy City's Hahn pumper, above, continues to pump water onto the rubble of the Davidson's Furniture Store the day following the fire. The destruction to this portion of the block is evident below. "Donut rolls" of hose have been piled on the tailboard of the engine, indicating that there is much more work to be done when the rig returns to the station. The blaze began in the kitchen of the J-A Restaurant and spread rapidly. Firefighters arrived to find flames shooting approximately 40 feet above the rear of the restaurant. (Above, Francis Senglar; below, HSSC.)

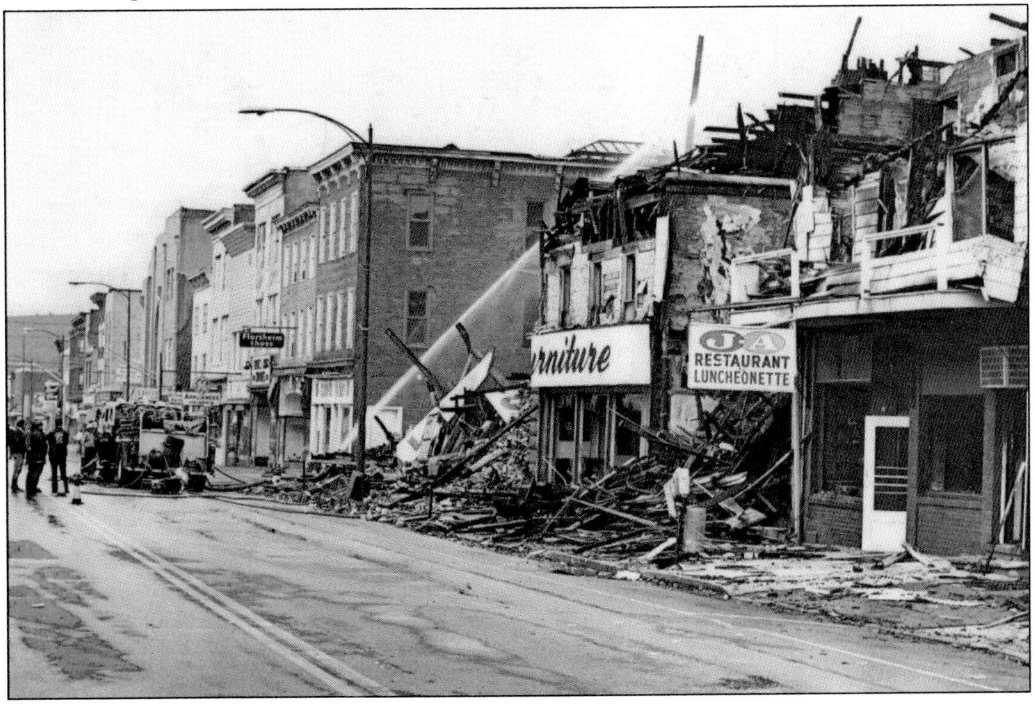

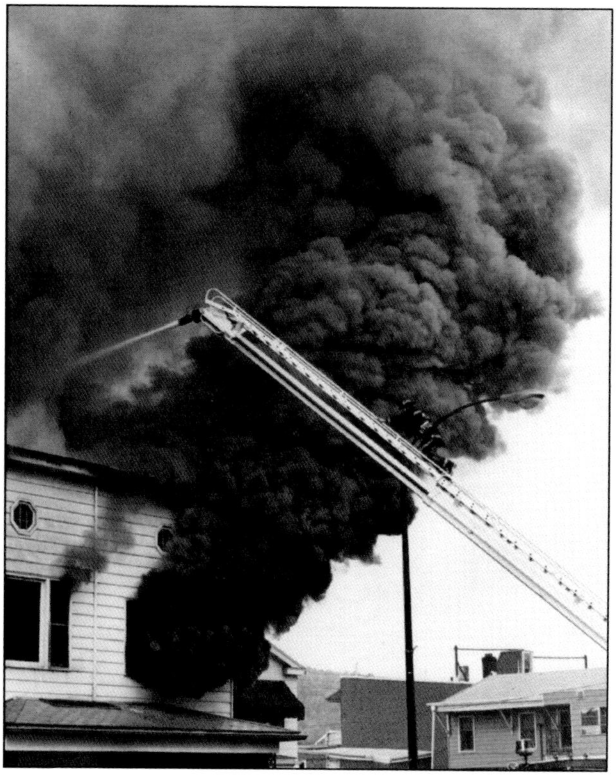

Inky smoke pours from the second-floor window of this end-of-the-row dwelling in Mahanoy City. Fire consumed much of the 200 block of East Centre Street on July 4, 1977. Shenandoah's American LaFrance TeleSQURT delivers an elevated master stream. Twenty individuals were left homeless, and seven firefighters were injured. The alarm was transmitted at 5:30 a.m. (HSSC.)

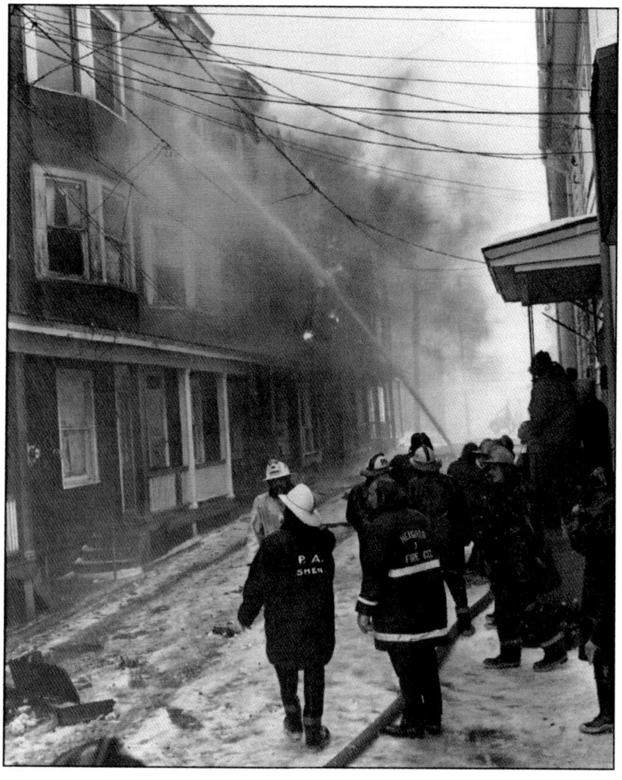

Firefighters direct a hose stream into the upper floors of this row of frame dwellings on a very narrow South Lehigh Street in Shenandoah on February 14, 1975. Seventeen homes were destroyed, and 23 people were left homeless by the fire. (HSSC.)

Washington Hook and Ladder of Mahanoy City swings into South Lehigh Street in Shenandoah with its American LaFrance aerial ladder truck. Frame row houses such as these typically featured an open area between the top floor ceiling and the roof deck known as a cockloft, which often extended the length of the row of homes. Once fire entered this area, it presented a huge challenge for firefighters. (HSSC.)

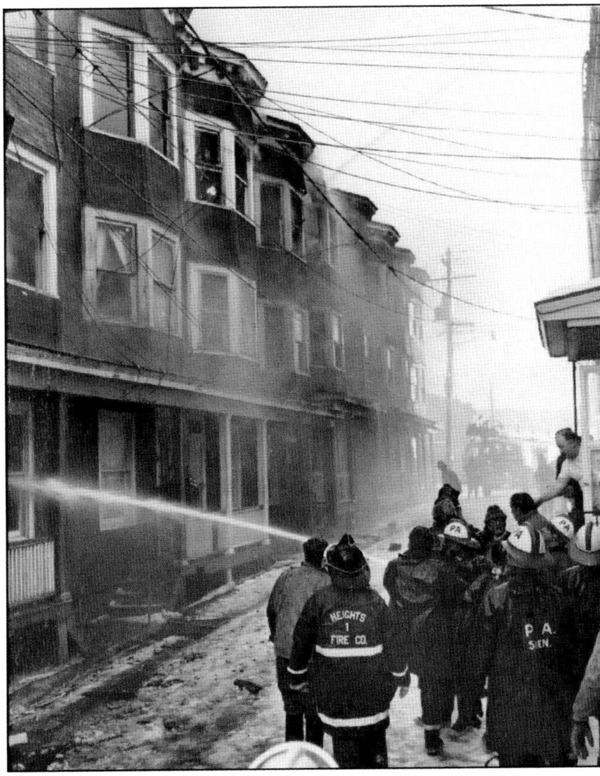

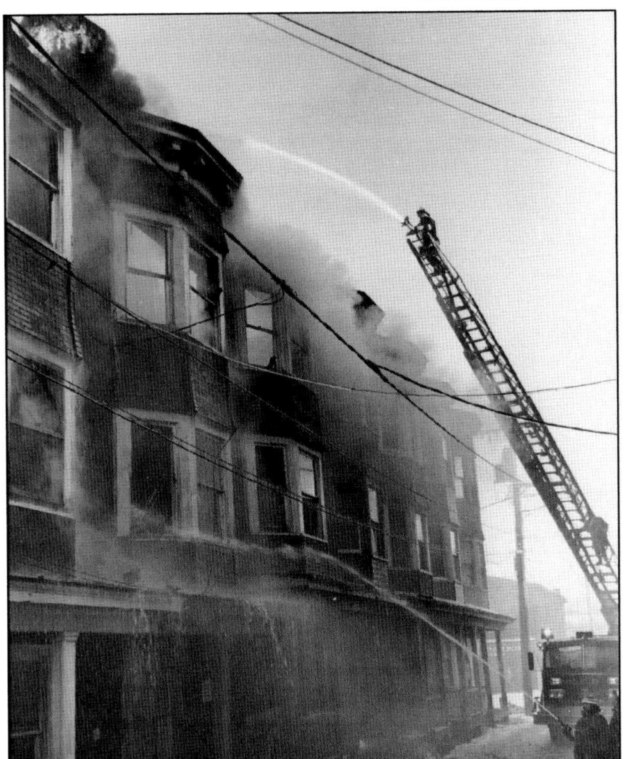

Mahanoy City's aerial ladder truck provides an elevated master stream to battle the fire on South Lehigh Street in Shenandoah. In addition to the five companies of the Shenandoah Fire Department, three Mahanoy City companies, Frackville, and Ringtown were on scene. Fire companies from Ashland, Girardville, and Pottsville were placed on standby. (HSSC.)

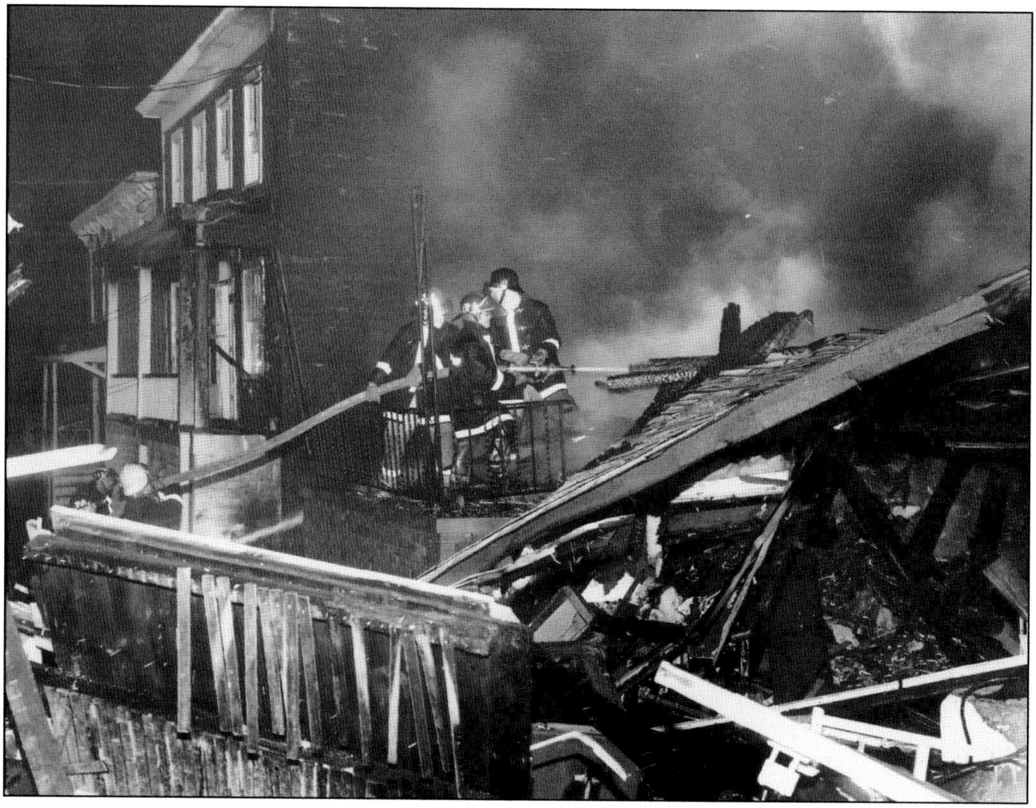

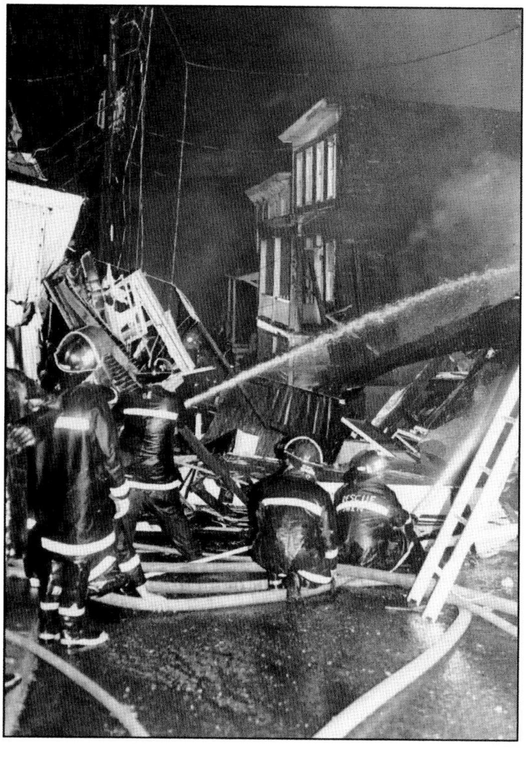

All that remains is the front porch from which firefighters direct a hose stream into the burning rubble in Shenandoah on February 1, 1983. A 66-year-old woman died in a 509 West Columbus Street home. The occupant of the home in which the explosion and fire originated was later arrested and charged with arson and homicide. (HSSC.)

Shenandoah firefighters pour water on the remains of three homes that were leveled by the blast on February 1, 1983. The alarm was sounded at 6:40 p.m. for an explosion and fire that originated in 507 West Columbus Street. (HSSC.)

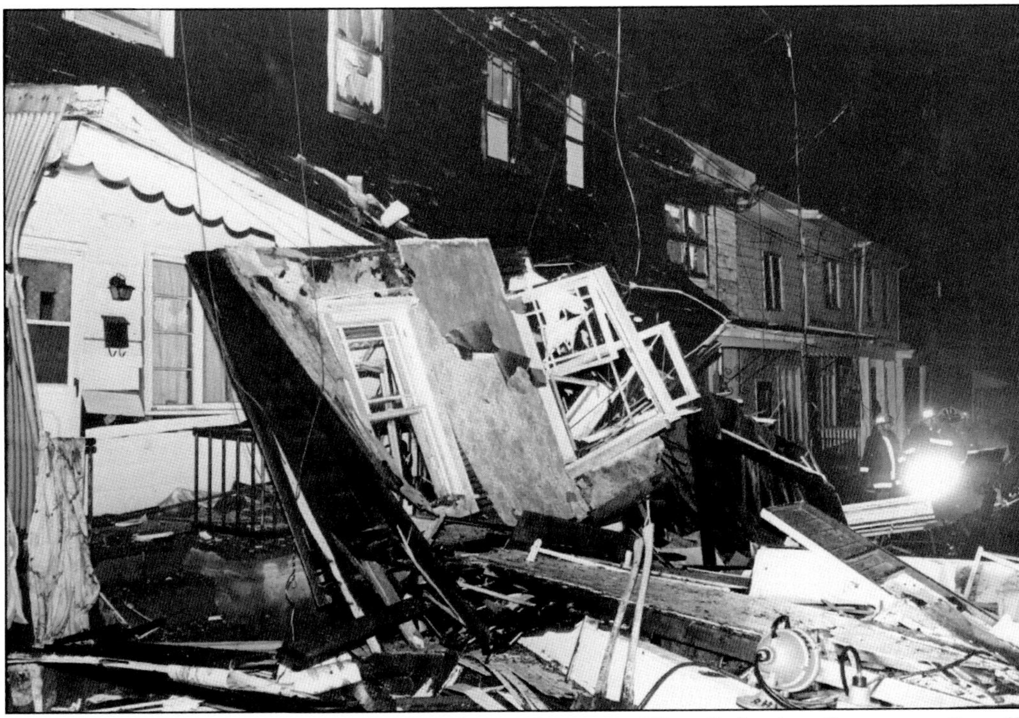

The front portions of homes involved in the explosion on West Columbus Street were literally blown across the street and rest against neighboring homes. Damage to those homes is evident as well. Two firefighters were injured contending with the flames. (HSSC.)

At 4:20 p.m. on November 20, 1989, Girardville firefighters were dispatched to 362 West Main Street for a fire on the second floor of a frame, middle-of-the-row home. Firefighters arrived to find an advanced fire that eventually extended to nine houses in the row. Numerous fire companies from the area assisted Girardville. The fire was caused by a defective baseboard electric heating unit. (Dan Markiewicz.)

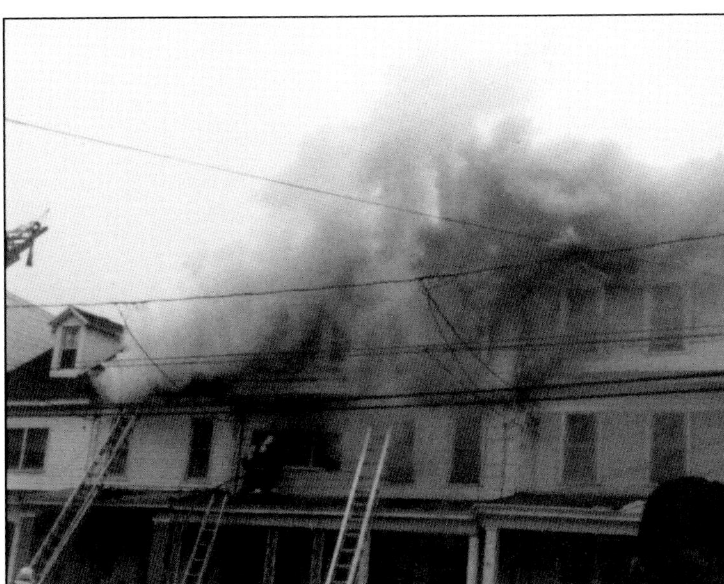

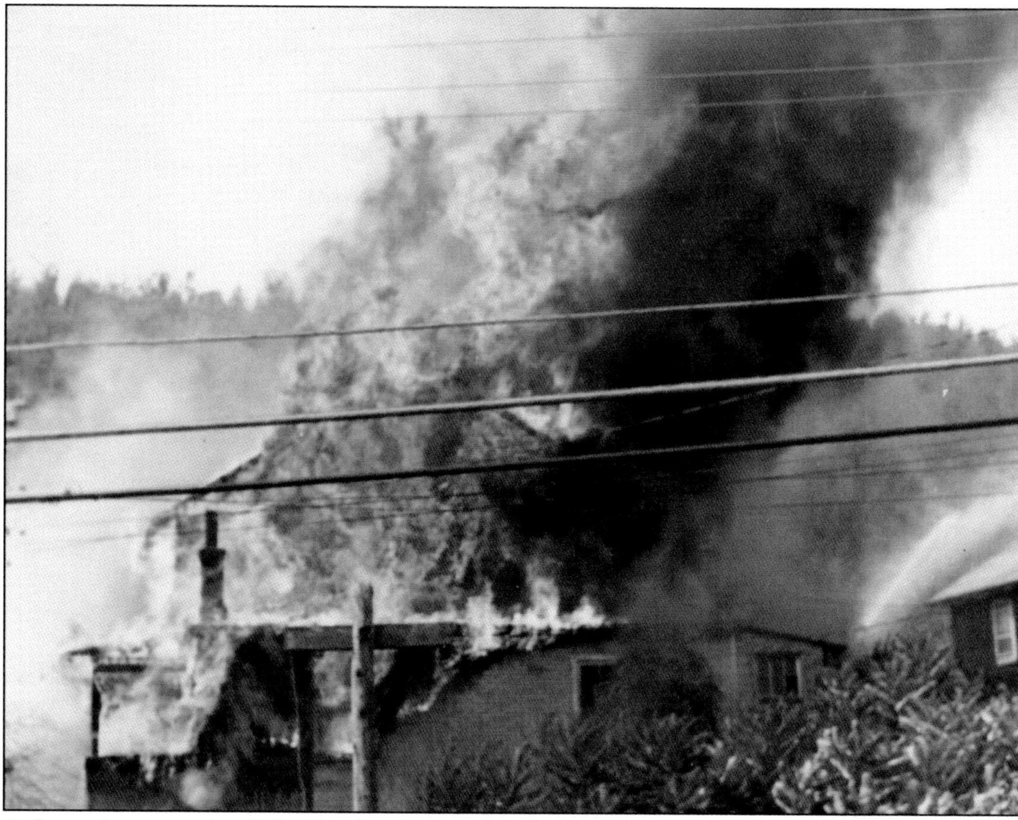

A frame home in the Coles Patch section of Mahanoy Township is engulfed in flames at this midday fire that occurred on Sunday, August 10, 1975. Four homes burned to the ground, two others were gutted, and two others heavily damaged leaving 15 people homeless. (HSSC.)

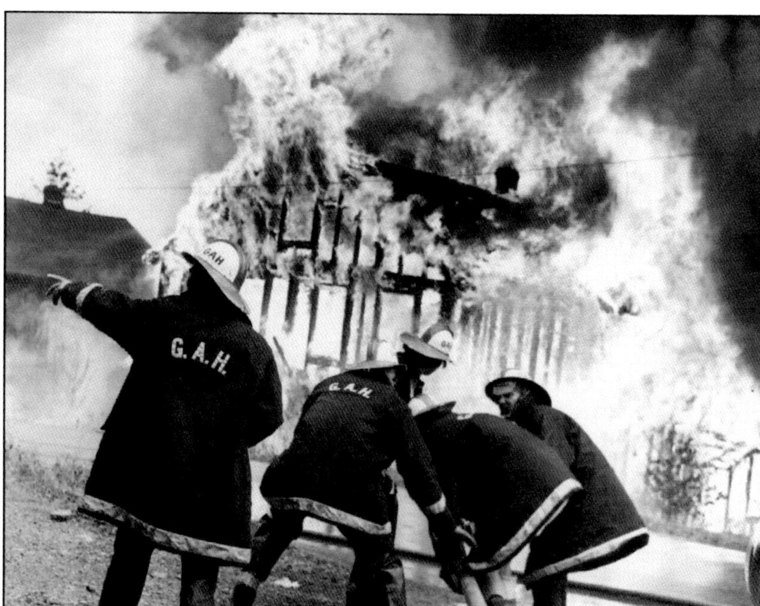

Firefighters from the Good American Hose Company of Mahanoy City direct a hose stream into a fully involved frame dwelling in Coles Patch on August 10, 1975. Four firefighters were injured fighting the fire, which began in a vacant frame duplex and quickly spread to surrounding homes as water supply problems hampered operations. (SHFS.)

Operating during a snowfall, a firefighter carries a saw to the roof at this fire in the 500 block of East Mahanoy Avenue in Mahanoy City on January 29, 1978. Another firefighter has a 2.5-inch hose line at the ready. The alarm was transmitted at 11:38 a.m. (HSSC.)

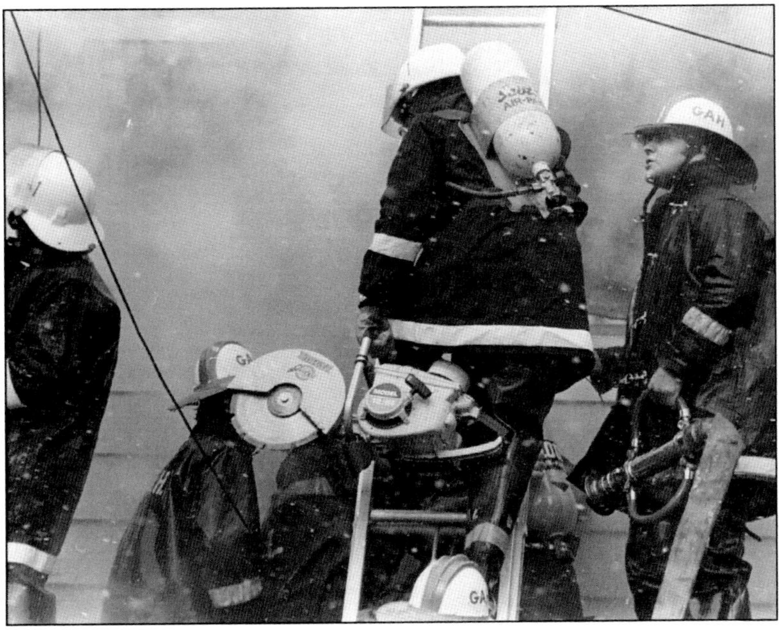

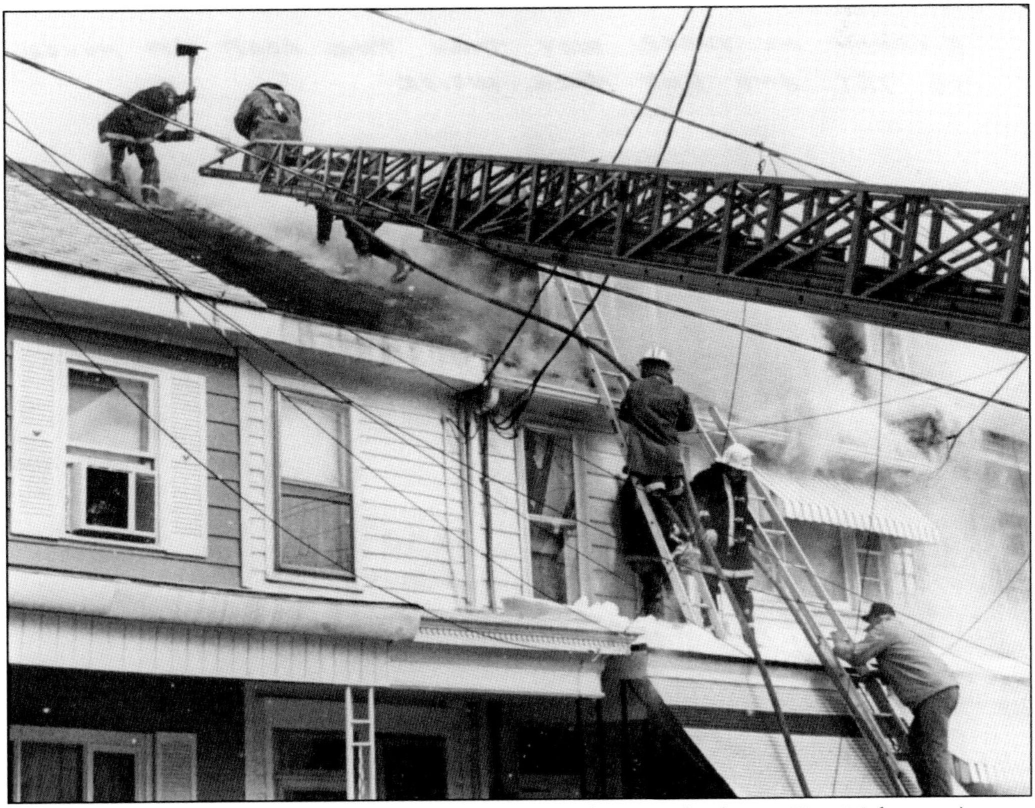

A firefighter with an axe vents a roof the old-fashioned way at the fire on East Mahanoy Avenue in Mahanoy City on January 29, 1978. Firemen are working to stop the spread of the fire through the attics of the frame dwellings. The Mahanoy Plane, Ryan Township, and Rainbow Hose Company of Schuylkill Haven assisted Mahanoy City. (HSSC.)

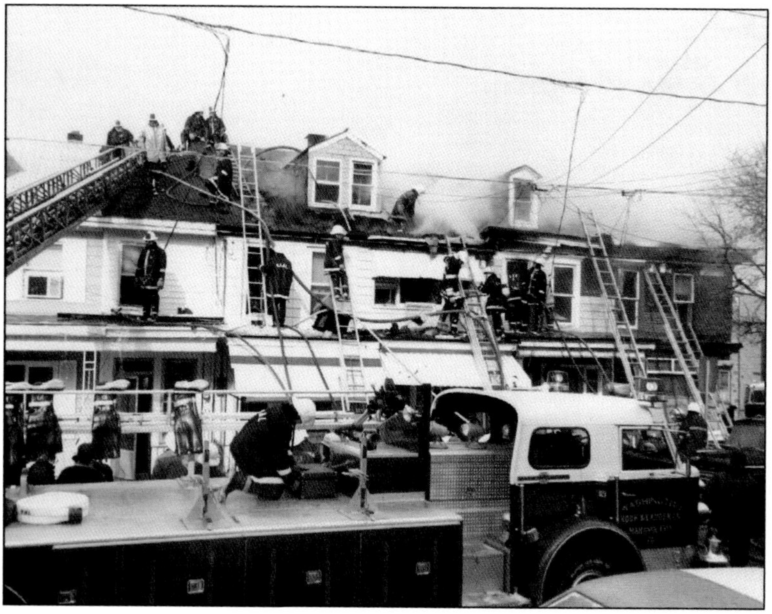

Having stopped the fire's progress, firefighters overhaul the buildings on East Mahanoy Avenue in order to extinguish any remaining pockets of fire. Three two-and-a-half-story homes were destroyed, and a café and two other homes were damaged. The fire was caused by a faulty chimney, which ignited the first-floor baseboards in one of the homes. (HSSC.)

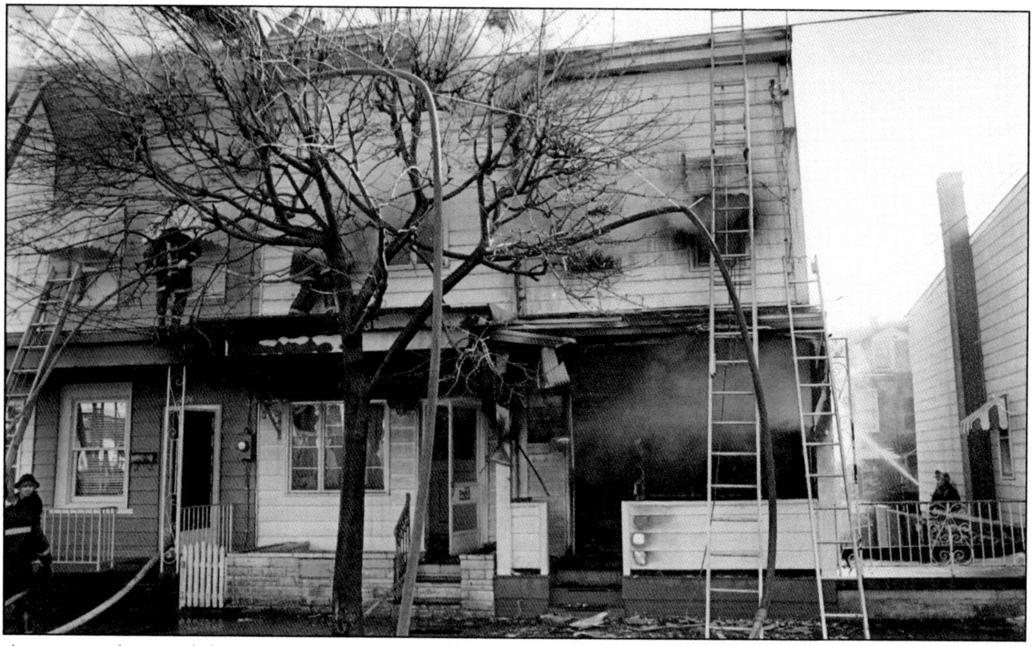

A routine home delivery of propane at 922 East Centre Street in Mahanoy City on January 9, 1982, turned deadly and destructive. The gas ignited and engulfed the end-of-the-row home. The female occupant of the home died as a result of burns sustained in the fire. Her husband was also seriously burned but survived. Hose lines lie draped over utility wires, through the trees, and to the roof. (HSSC.)

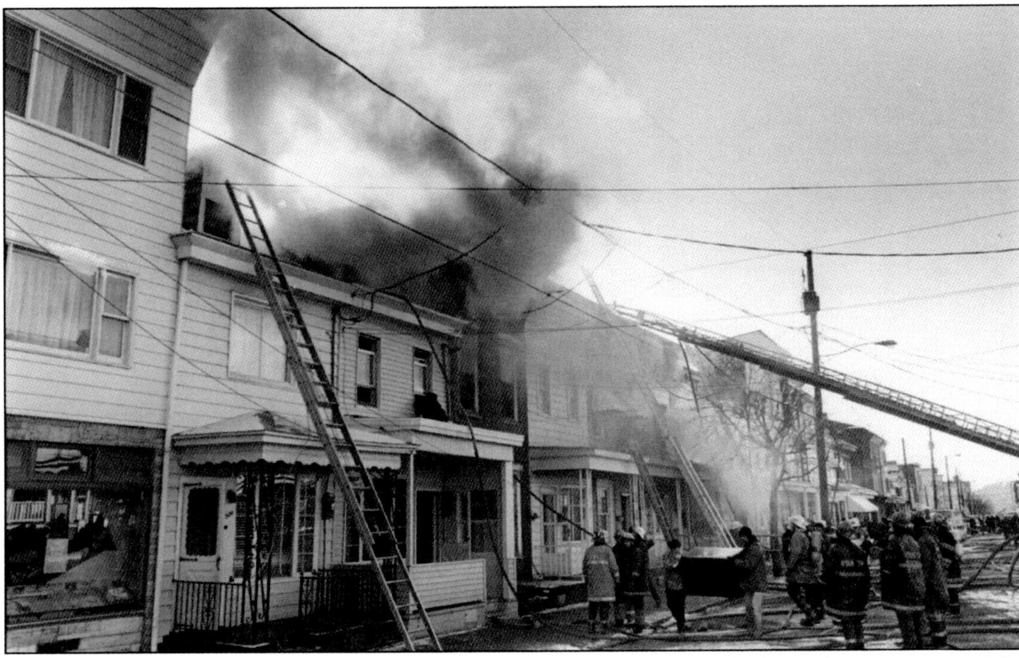

As firefighters work to stop the spread of the fire through the cocklofts and attic spaces of the homes on East Centre Street in Mahanoy City, individuals remove furniture from the fire's path. The alarm was received at 1:10 p.m. In all, the fire destroyed or damaged five homes and left 14 people homeless. (HSSC.)

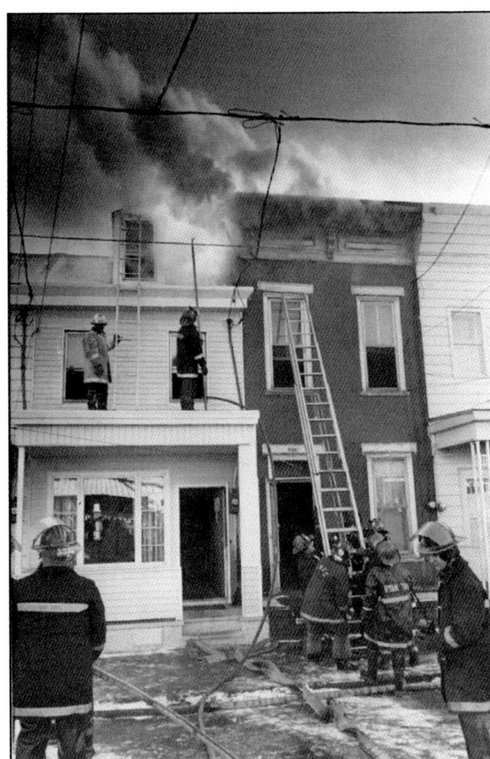

With heavy smoke from the eaves indicating fire running the cockloft, firefighters place a 35-foot ladder to this home in the 900 block of East Centre Street on January 9, 1982. Mahanoy City fire chief Paul MacLeary requested mutual aid from several surrounding fire departments in order to battle the fire in freezing temperatures. Two firefighters were injured, one of whom was hospitalized, while fighting the fire. (HSSC.)

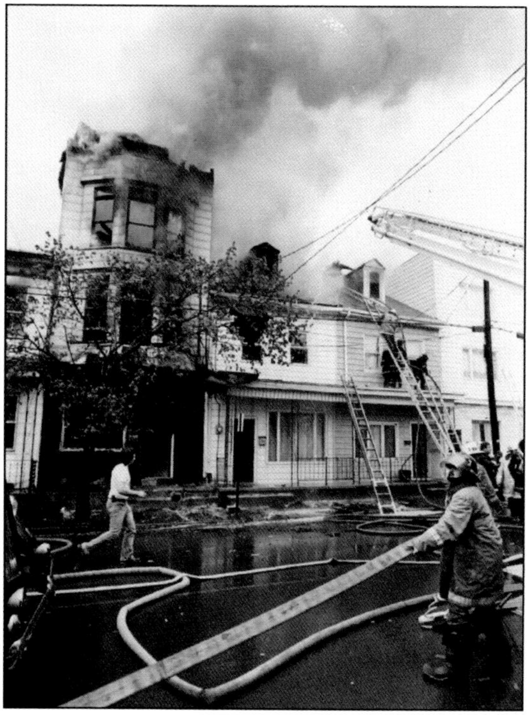

Heavy fire was showing from the three-story frame dwelling at 507 East Center Street when firefighters arrived in response to the 11:12 a.m. alarm in Shenandoah on May 4, 1983. The fire extended to frame dwellings on both sides of 507, destroying or damaging six and leaving 17 individuals homeless. (HSSC.)

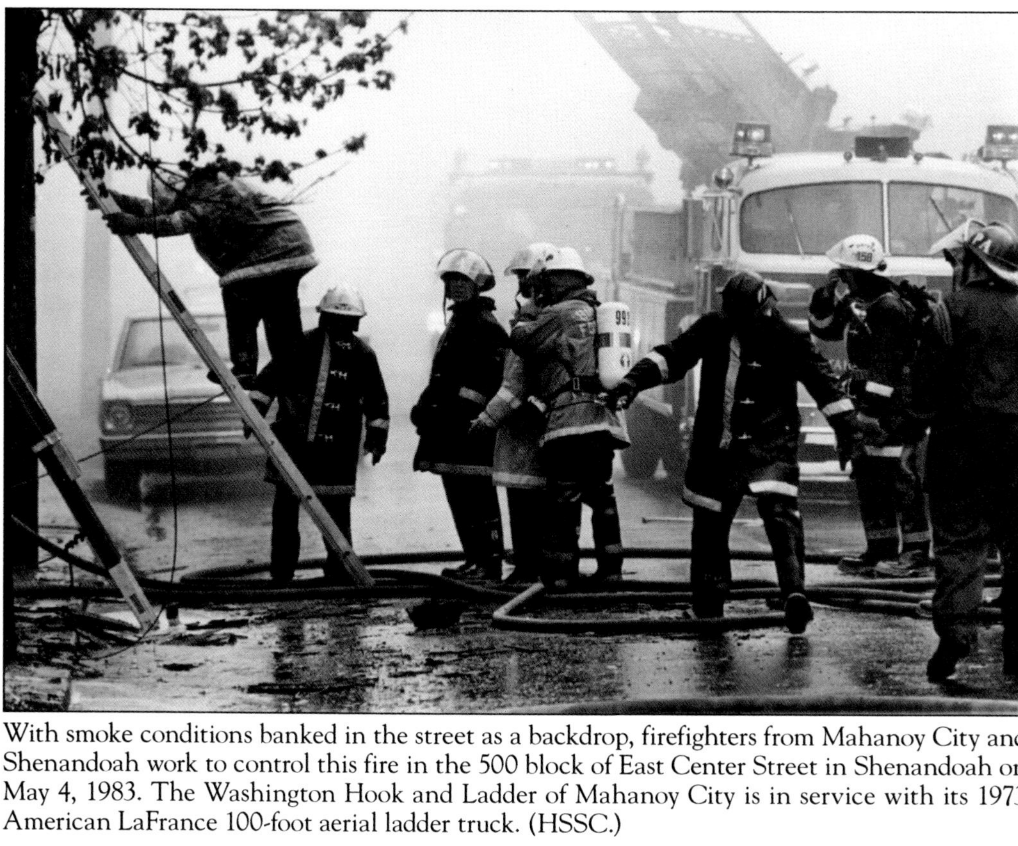

With smoke conditions banked in the street as a backdrop, firefighters from Mahanoy City and Shenandoah work to control this fire in the 500 block of East Center Street in Shenandoah on May 4, 1983. The Washington Hook and Ladder of Mahanoy City is in service with its 1973 American LaFrance 100-foot aerial ladder truck. (HSSC.)

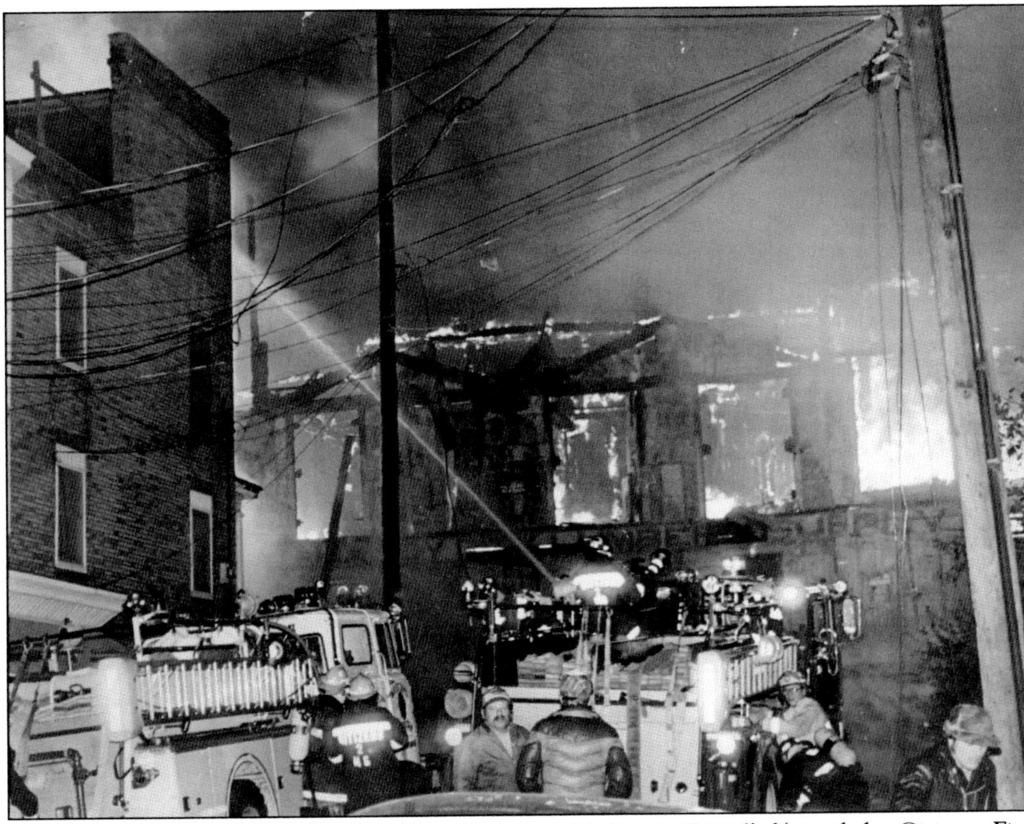

Mack CF pumpers of the West End Fire Company, Mahanoy City (left), and the Citizens Fire Company, Mahanoy City (right), operate side by side in battling this fire at the warehouse of the Mahanoy City Lumber and Supply Company on November 17, 1984. The alarm was sounded at 2:40 a.m. Firefighters arrived to find fire extending 400 feet through the structures along Railroad Street. Both pumpers were damaged by radiant heat. (Dan Markiewicz.)

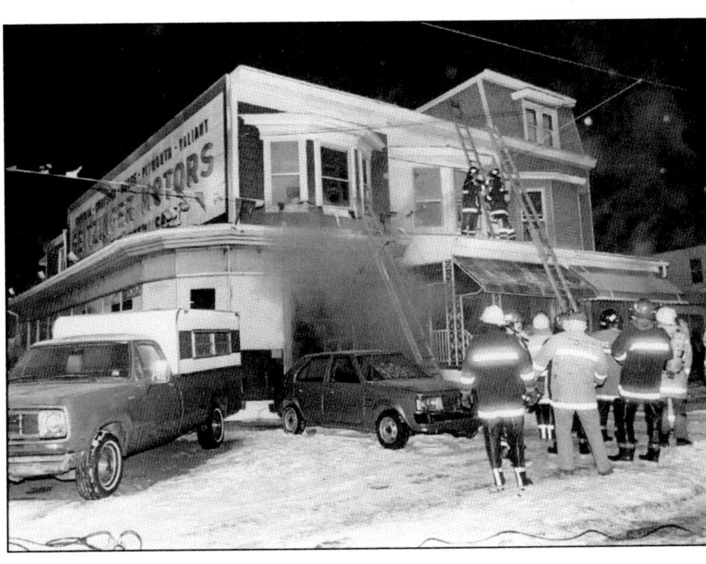

At 8:12 p.m. on Christmas Day, 1983, firefighters from the Frackville Fire Department were dispatched to Seitzinger Motors at 450 South Lehigh Avenue for a fire that caused $50,000 damage to the business and resulted in smoke damage to two adjoining homes. The fire was caused by a malfunctioning supplemental heater. The temperature at the time of the fire was -2 degrees. (HSSC.)

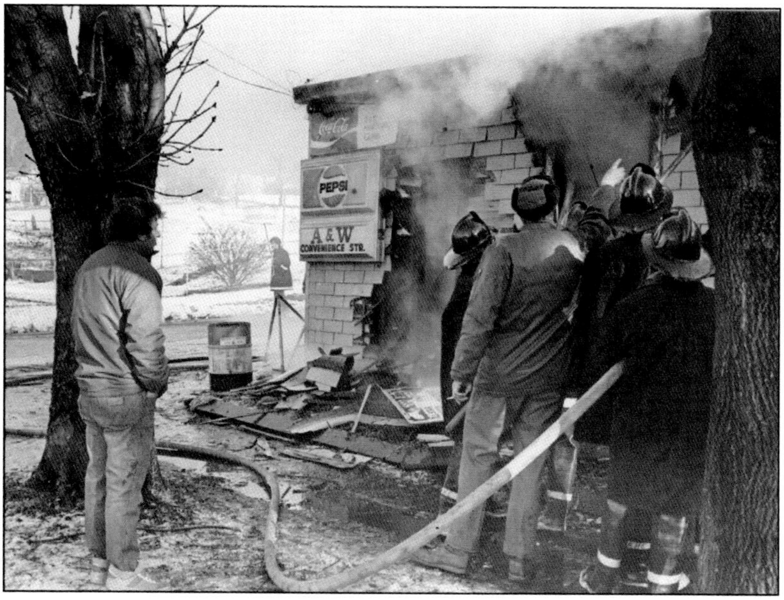

Firefighters are overhauling this fire at the A and W Convenience Store in Fountain Springs on December 29, 1978. The alarm was sounded at 9:00 a.m. Fire companies from Fountain Springs, Ashland, and Gordon responded. (HSSC.)

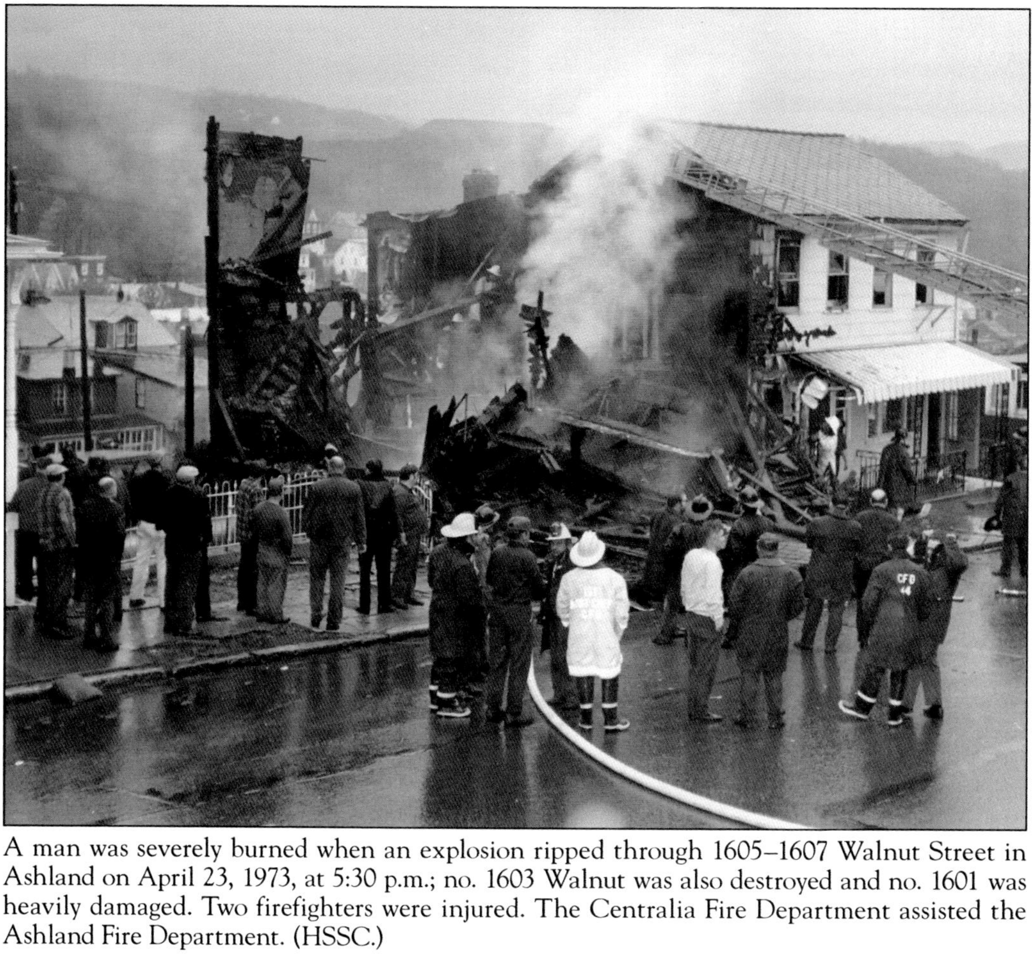

A man was severely burned when an explosion ripped through 1605–1607 Walnut Street in Ashland on April 23, 1973, at 5:30 p.m.; no. 1603 Walnut was also destroyed and no. 1601 was heavily damaged. Two firefighters were injured. The Centralia Fire Department assisted the Ashland Fire Department. (HSSC.)

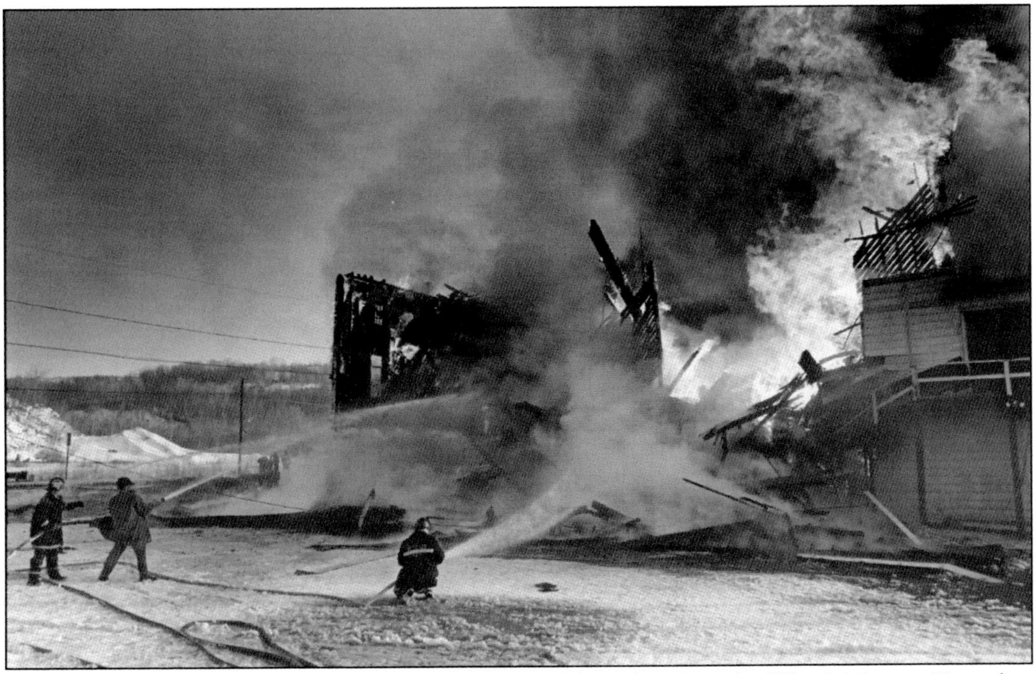

On January 19, 1984, fire destroyed the St. Mary Magdalen Church in the West Mahanoy Township village of Lost Creek. In the photograph, fire consumes the sanctuary as much of the structure begins to collapse. The alarm was transmitted at 3:00 p.m. The church was built in 1939. While the sanctuary was destroyed, firefighters saved the rectory. (HSSC.)

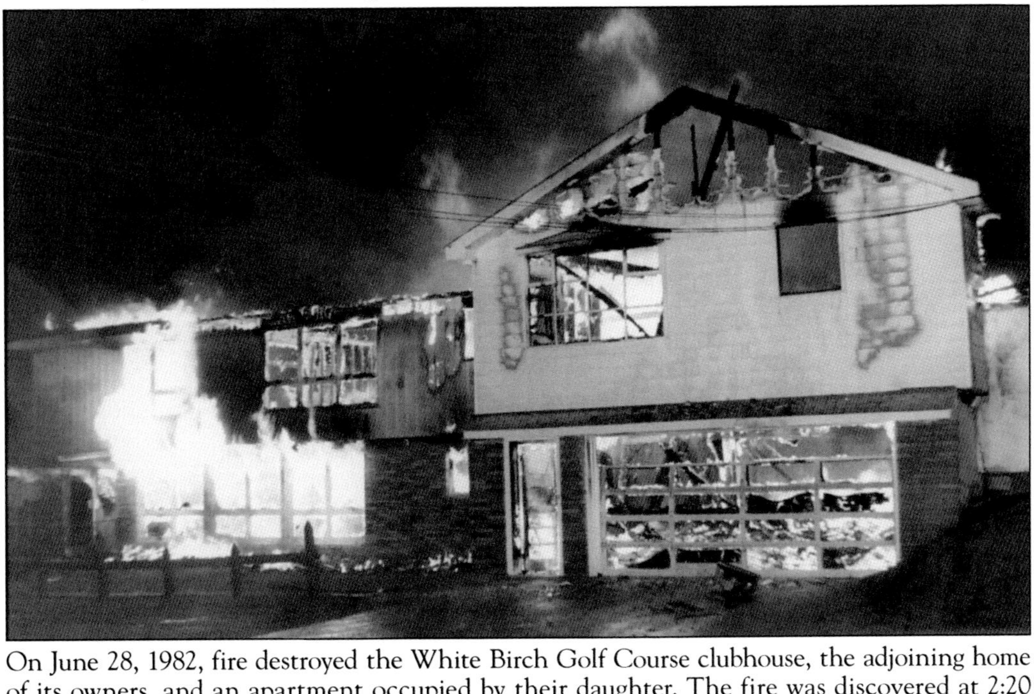

On June 28, 1982, fire destroyed the White Birch Golf Course clubhouse, the adjoining home of its owners, and an apartment occupied by their daughter. The fire was discovered at 2:20 a.m. Firefighters from Ryan Township, Hometown, Quakake, New England, and Mahanoy City responded to the fire, which could be seen for miles. One firefighter was hospitalized. (HSSC.)

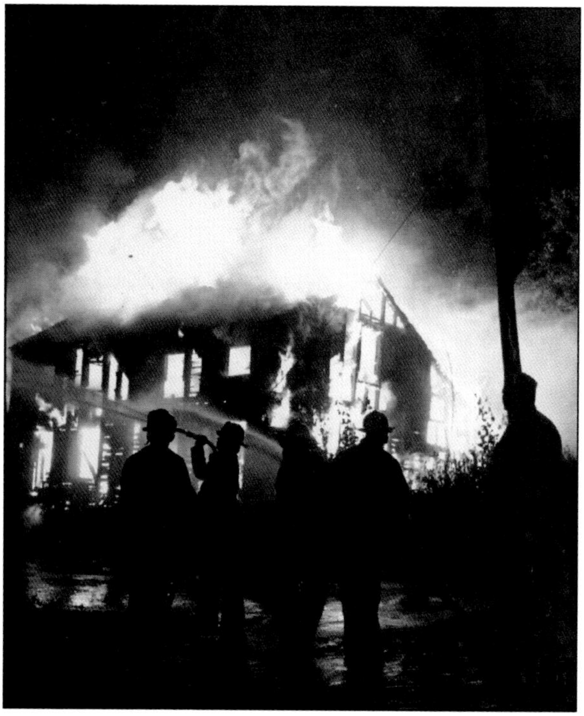

At 2 a.m. on Sunday, July 25, 1976, Ashland firefighters arrived to find this vacant home completely engulfed in flames. Firefighters worked to protect nearby buildings that were threatened by radiant heat. The original fire building eventually collapsed. (HSSC.)

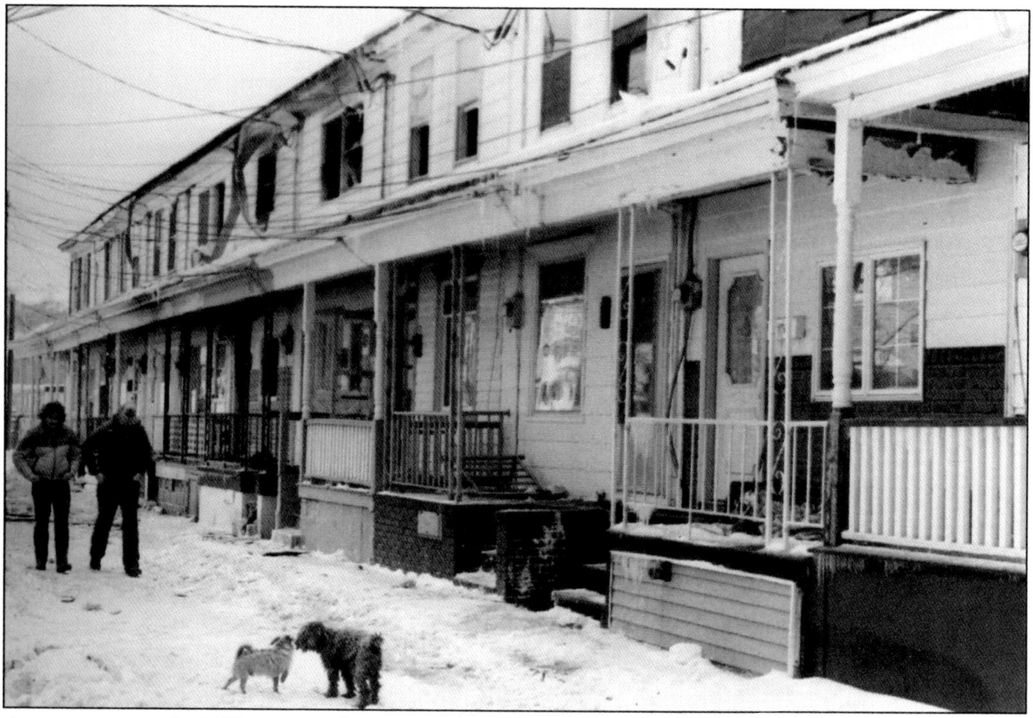

At approximately 1:20 a.m. on January 24, 1982, fire began in the kitchen of 139 C Street in Girardville and spread rapidly through 10 frame row houses. Twenty-eight individuals were left homeless by the fire, and seven firefighters were injured. The snow and ice visible the day following the fire is evidence of the conditions that firefighters faced in battling the flames. (HSSC.)

Looking down the 1900 block of Center Street in Ashland on January 24, 1981, fire is spreading rapidly through the frame dwellings. Just to the right of the utility pole in the foreground, the Washington Fire Company can be seen raising the 85-foot aerial on its 1959 American LaFrance ladder truck. All the buildings pictured in the row were destroyed in the fire. (American Fire Company, Fountain Springs.)

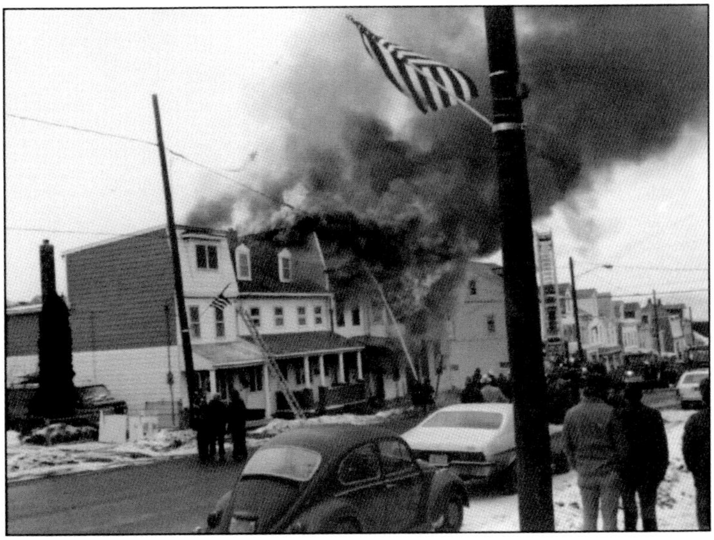

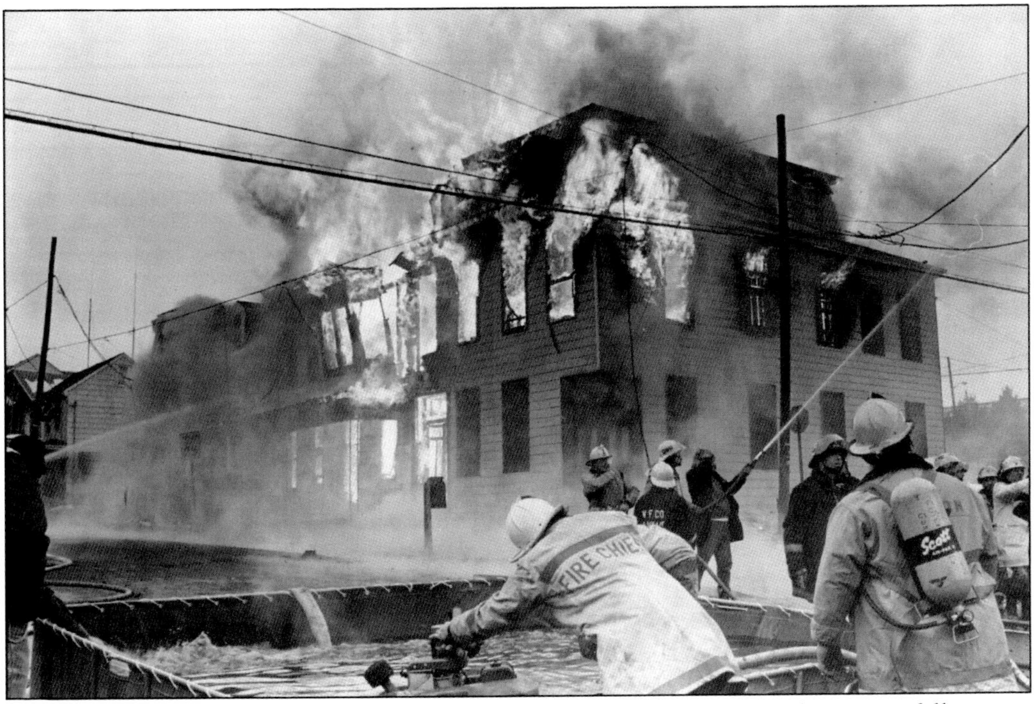

In November 1980, Ashland firefighters, faced with a depleted municipal reservoir following a prolonged drought, developed contingency plans for an alternate water supply in the event of a fire in the borough. Shortly after implementation, the plan was tested on January 24, 1981, at 9:00 a.m. when fire struck the 1900 block of Center Street, killing a 24-year-old mother and her four-year-old child. (HSSC.)

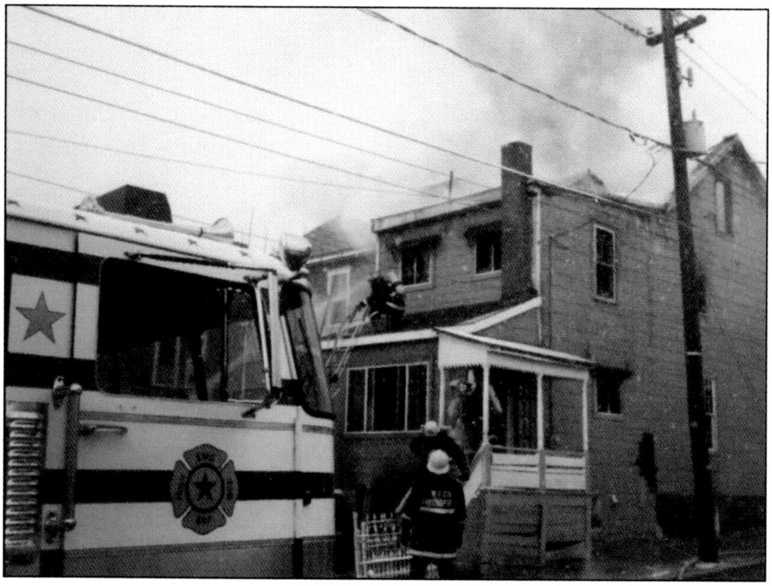

Rangers Hose Company of Girardville's 1974 Hahn pumper is in service at this fire in April 1990, which damaged several frame homes on Walnut Street in Ashland. The Rangers' Hahn sports the trademark red, white, and blue color scheme, which made the company's apparatus unique for many years. (Authors' collection.)

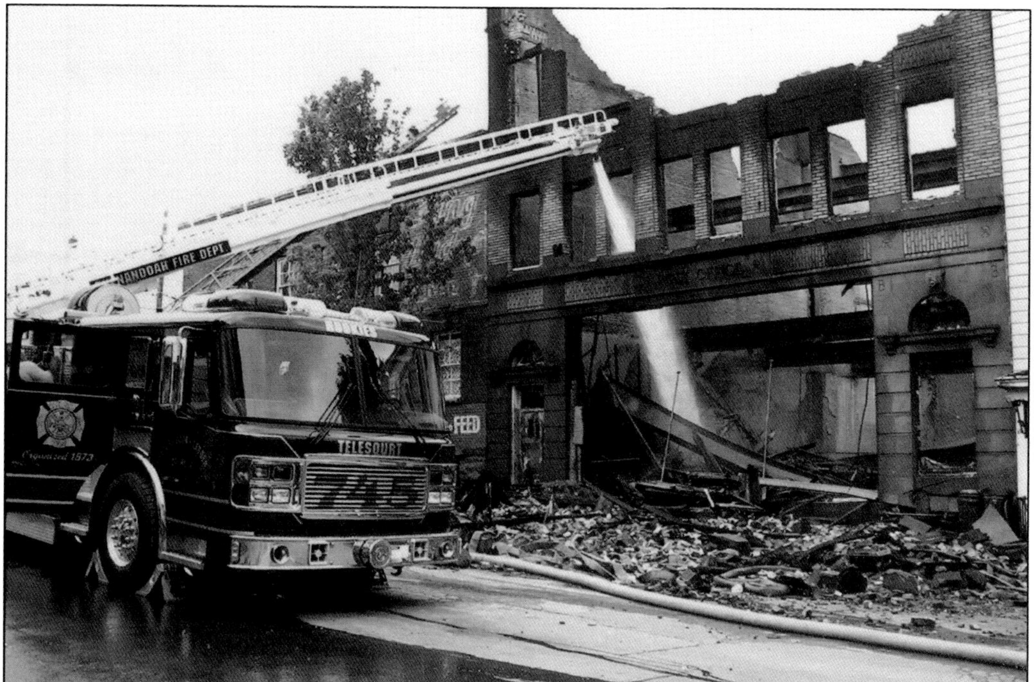

Rescue Hook and Ladder of Shenandoah operates a new 2007 American LaFrance 50-foot TeleSQURT at this fire that destroyed the historic building that once housed the Shenandoah Post Office. The June 10, 2007, fire on South Main Street originated in the alley to the rear in a motor home and quickly spread to adjacent frame structures. (Dan Markiewicz.)

Two

SOUTH

Organized firefighting commenced in Schuylkill County in the early 19th century. Residents in the thriving farming community of Orwigsburg established the Orwigsburg Fire Company, later renamed Friendship Hose Company, about 1809. Friendship Hose Company's Pat Lyon hand pumper, built in 1809, represents Schuylkill County's first fire apparatus.

With the opening of the Schuylkill Canal in 1825, Schuylkill County's river towns began to develop. Schuylkill Haven, southern Schuylkill County's largest community, initiated a fire department in 1885 with the formation of the Good Intent Fire Company No. 1, which was later renamed the Spring Garden Hose Company before finally being called the Rainbow Hose Company. Also joining the department was the Friendship Hook and Ladder, which merged with the Rainbow Hose Company in 1908. Additional Schuylkill Haven fire companies included the Schuylkill Hose Company, established in 1892, and the Liberty Fire Company, organized in 1908.

Another river community, Auburn, assembled its Vigilant Fire Company No. 1 in 1898. This company was later called the Auburn Fire Company. The borough of Cressona added two fire companies, the Cressona Fire Company No. 1 and the Good Will Fire Company, in 1899 and 1910, respectively. Further along the Schuylkill River near the Berks County border, Port Clinton started a volunteer fire company in 1946. Another small farming town, New Ringgold, formed the Community Fire Company in 1926.

Farming communities realized the importance of inaugurating volunteer fire companies for protection. Prior to the Civil War, Pine Grove put together its Hose, Hook, and Ladder Company in 1855 and a second company, the North End, in 1935. Pine Grove Township formed two companies in the villages of Ravine and Suedburg in 1946 and 1948, respectively. The tiny boroughs of Deer Lake and Landingville also developed fire companies about this time. The Deer Lake and West Brunswick Township Fire Company organized in 1944, and the Community Fire Company of Landingville formed in 1950. Another farming area, Wayne Township, activated two fire companies in the mid-20th century. The Summit Station Fire Company started in 1948, and the Friedensburg Fire Company formed later in 1960.

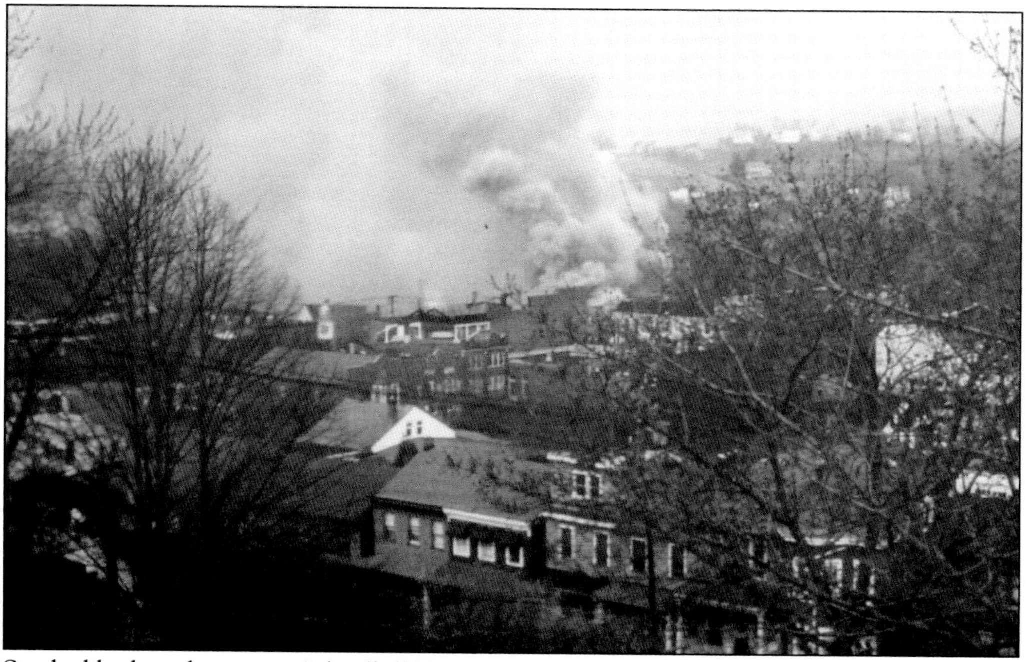

Smoke blankets downtown Schuylkill Haven on Easter Sunday morning, April 17, 1960, as fire heavily damages much of the unit block of West Main Street. Fire companies from Pottsville, Orwigsburg, and Cressona assisted the Schuylkill Haven Fire Department. (R. Nagle, www.schuylkillhavenhistory.com.)

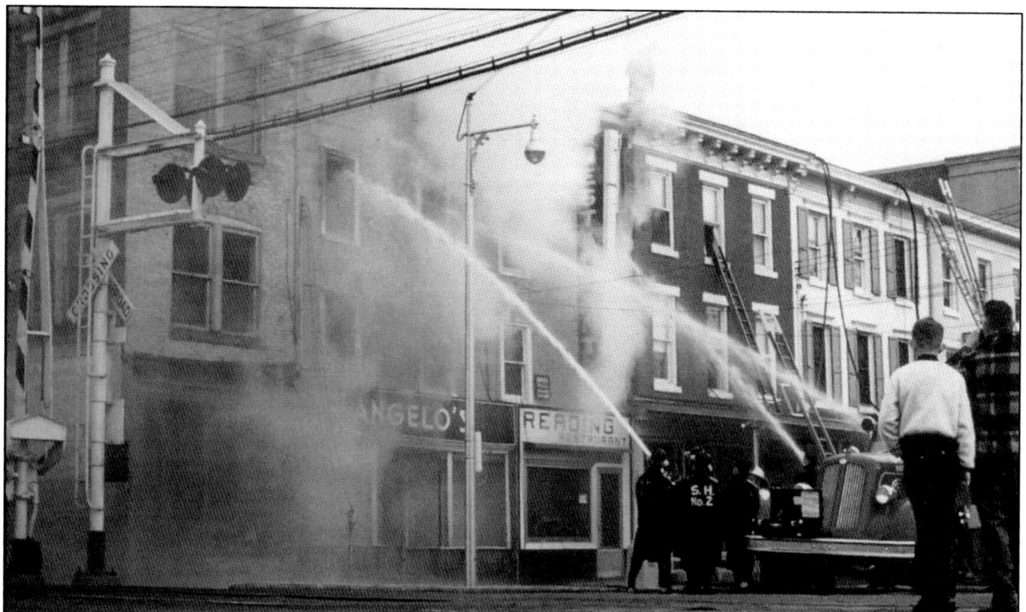

Rainbow Hose Company's classic Seagrave city service truck operates on West Main Street as ladders are placed to the exposure buildings and hose lines stretched to the roofs. An open self-contained breathing apparatus (SCBA) case lies on the front bumper of Rainbow's rig. A child watching the action on Easter morning, April 17, 1960, holds what appears to be a camera or camera case. (R. Nagle, www.schuylkillhavenhistory.com.)

Heavy smoke pours from Angelo's Town Tavern as well as the Palsgrove Hardware Store on the corner at 13 West Main Street. Spectators postpone Easter morning plans to watch firefighters battle the blaze. The fire was discovered shortly after 6:00 a.m. The losses totaled $75,000. (R. Nagle, www.schuylkillhavenhistory.com.)

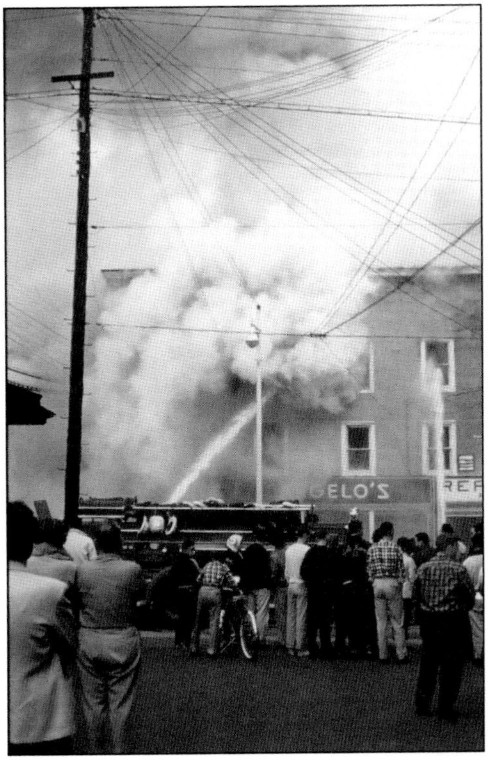

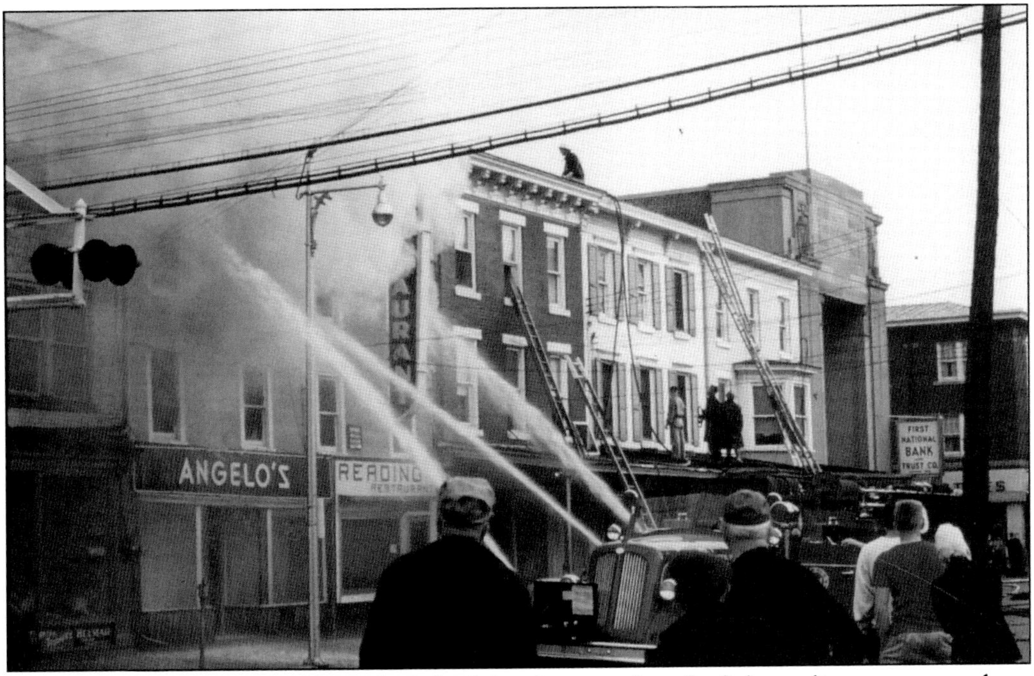

The railroad crossing is visible to the left of the photograph as firefighters direct numerous hose streams into both Angelo's Town Tavern at 11 West Main Street and the Reading Restaurant at 9 West Main Street. The fire of unknown cause was believed to have begun in the apartment above Angelo's. (R. Nagle, www.schuylkillhavenhistory.com.)

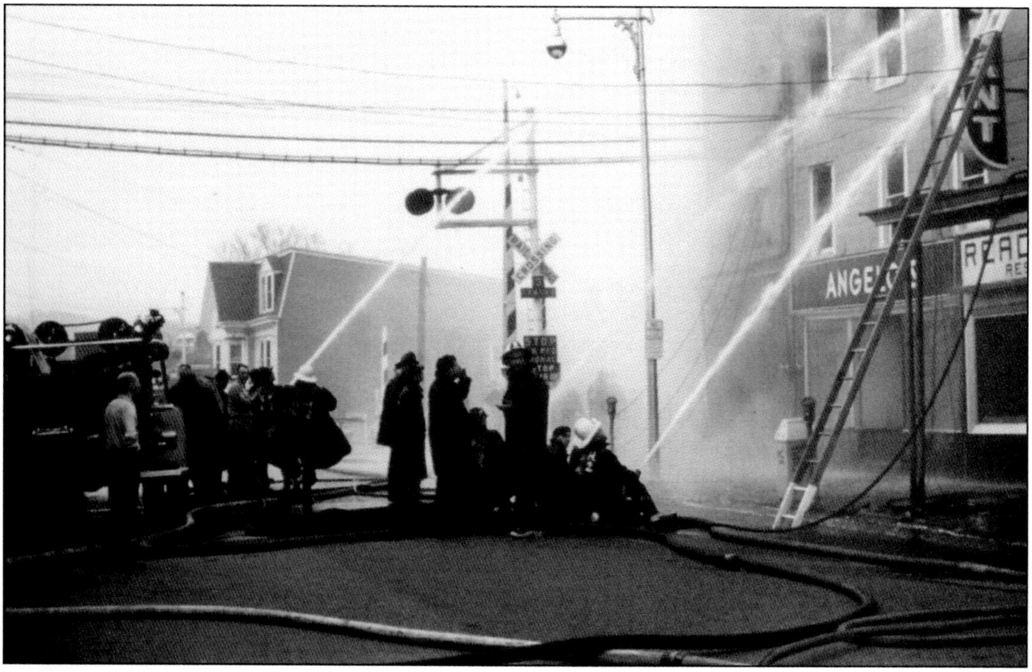

Firefighters have 2.5-inch hose lines placed in the street and are set up for a prolonged defensive struggle as fire burns in the buildings on West Main Street. The rear of Rainbow's Seagrave city service truck is visible to the left, with much of its compliment of ground ladders removed and placed in service. (R. Nagle, www.schuylkillhavenhistory.com.)

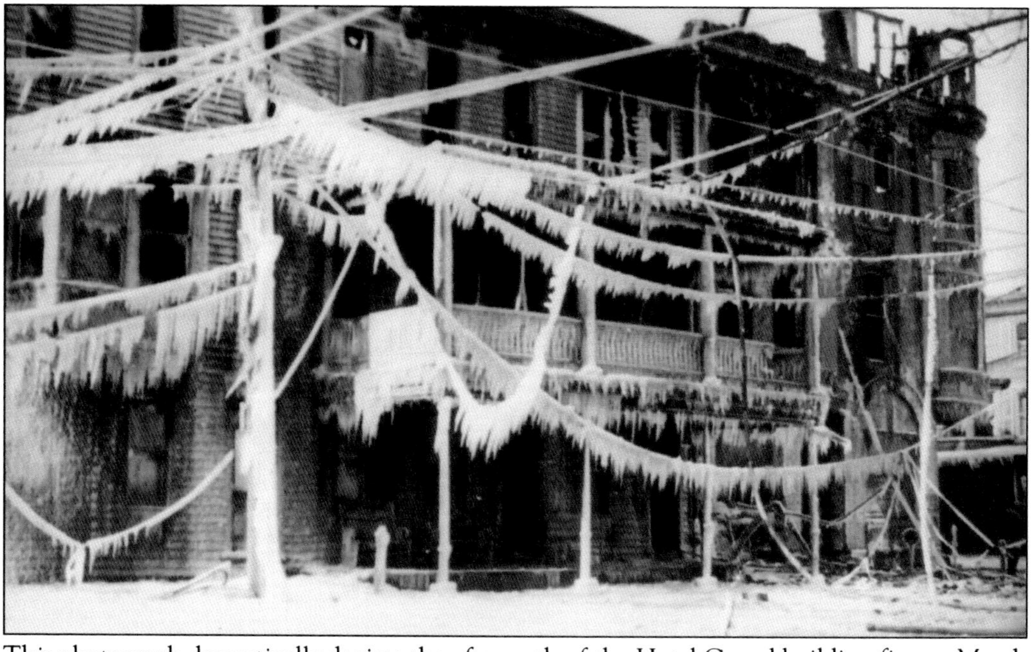

This photograph dramatically depicts the aftermath of the Hotel Grand building fire on March 8, 1932, in Schuylkill Haven. The ice accumulation is testament to a long battle with the flames, which resulted in a $60,000 loss. The building at Main and St. John Streets housed the post office and the Schuylkill Haven State Bank. (R. Nagle, www.schuylkillhavenhistory.com.)

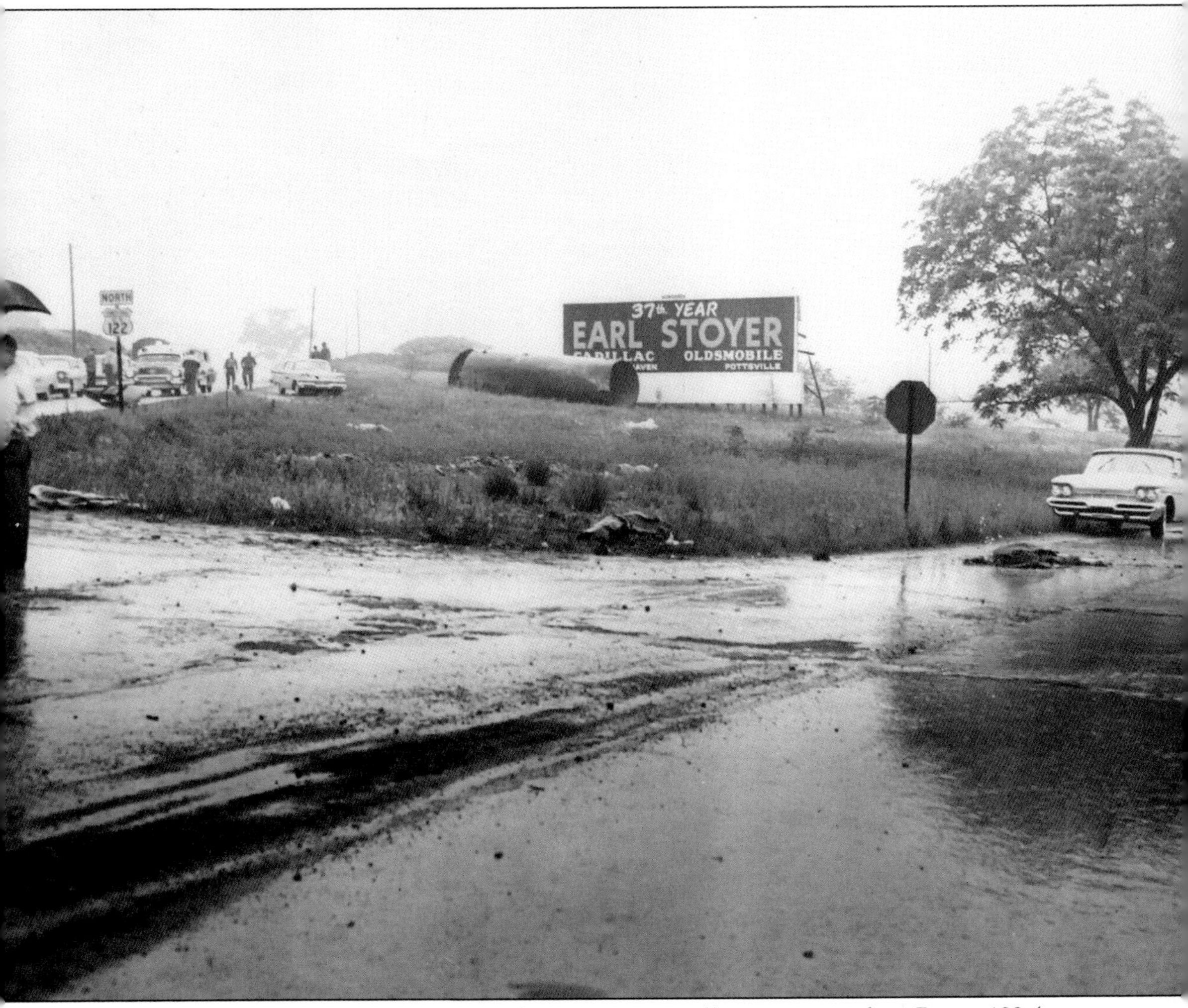

On Tuesday, June 2, 1959, at approximately 8:00 a.m., a collision occurred on Route 122 (now Route 61) just north of Deer Lake near the Red Church, when a Branch freight truck struck the rear of a Sun Gas Company propane tanker. The ensuing fire engulfed the propane tanker. The result was a BLEVE—Boiling Liquid, Expanding Vapor Explosion. The remains of the tank can be seen near the Earl Stoyer billboard several hundred feet north of the crash site. Ambulances can be seen on Route 122 to the left in the photograph. Sadly, bodies and body parts were strewn about this area as 11 individuals were killed by the explosion and resulting shrapnel. Some are seen covered with tarps. (HSSC.)

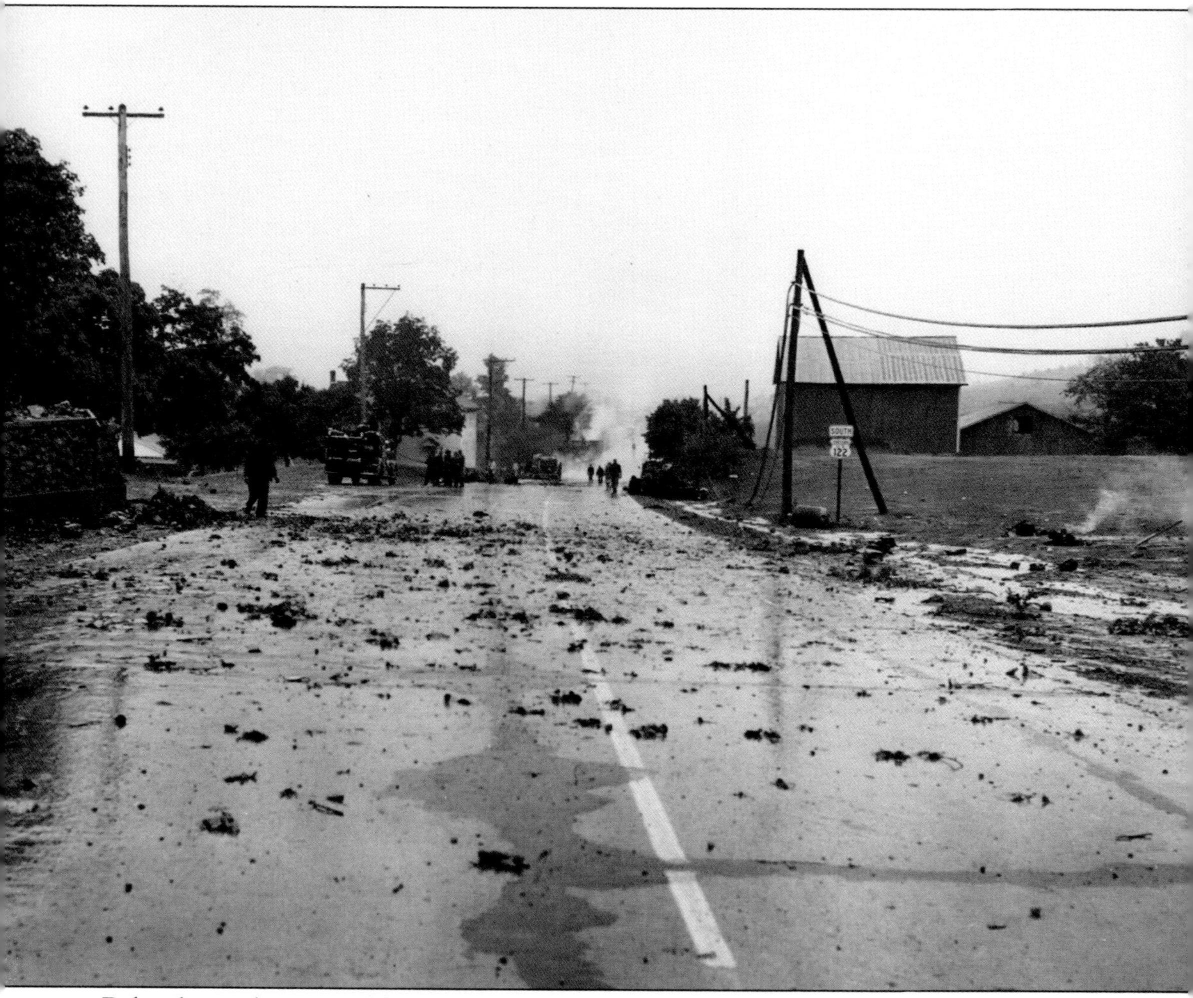

Debris litters the scene of the crash following the explosion of the propane tanker approximately 45 minutes after the initial crash on June 2, 1959. The mangled chassis of the truck can be seen on the right shoulder of the road near the downed utility wires. The rear of the Rainbow Hose Company of Schuylkill Haven's Seagrave city service ladder truck can be seen on the left shoulder. Eleven individuals were killed and 11 others injured when the explosion sent the tank and parts of the tank rocketing northward as seen in the photograph on page 45. The stone wall of the Red Church to the left was seriously damaged by the blast and bears scars to this day. (HSSC.)

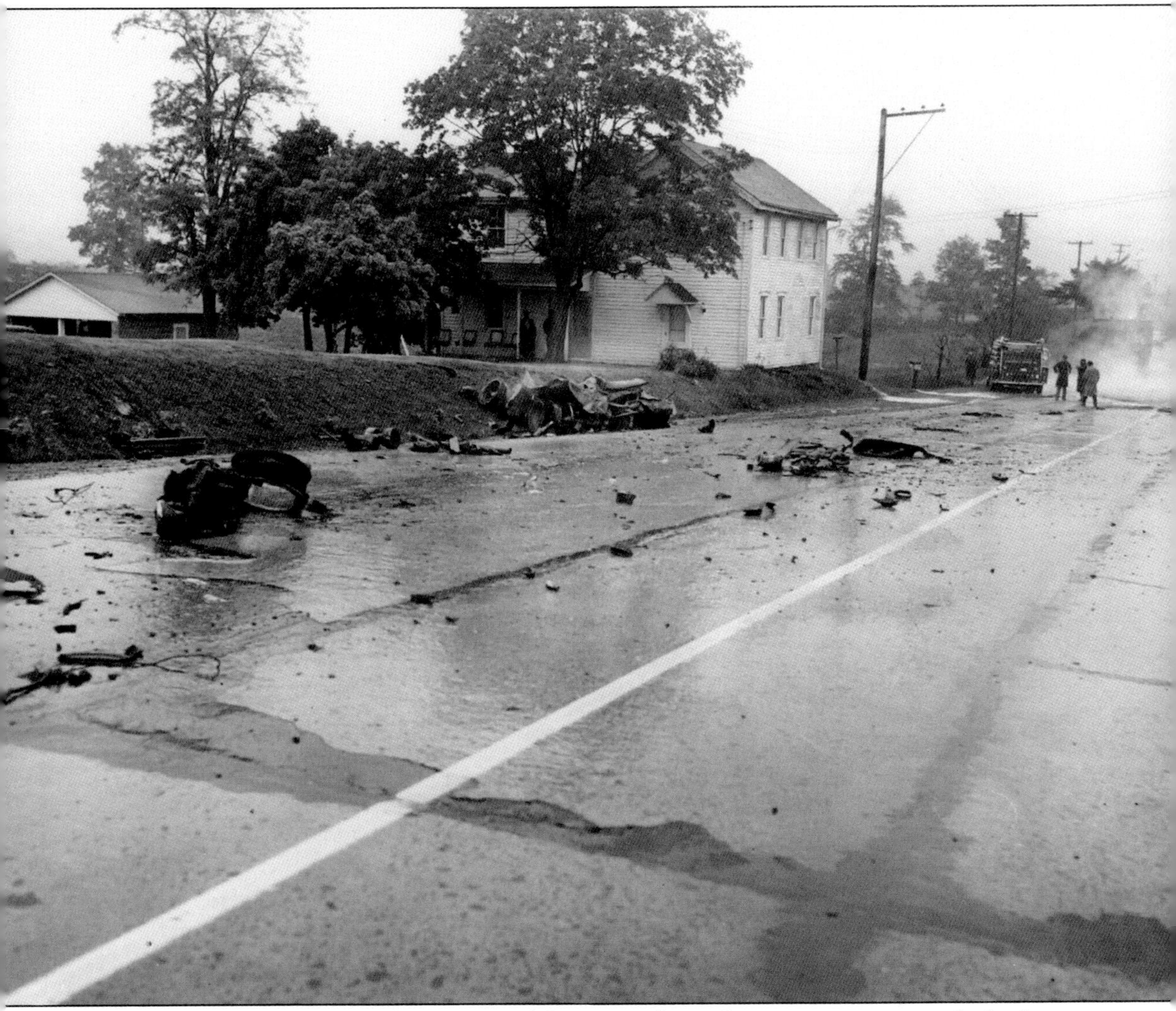

More pieces of the chassis of the propane delivery truck can be seen strewn over the highway following the explosion. The rear of Liberty Fire Company of Schuylkill Haven's 1957 Mack pumper can be seen on the highway as the Branch delivery truck smolders just to the right. The propane truck was stopped on the highway while a school bus in front of it picked up students when it was struck from behind by the freight truck. Thankfully the school bus was well clear of the scene when the explosion occurred and no students were injured. Unbelievably, the home on the left sustained only minor damage and still stands today. (HSSC.)

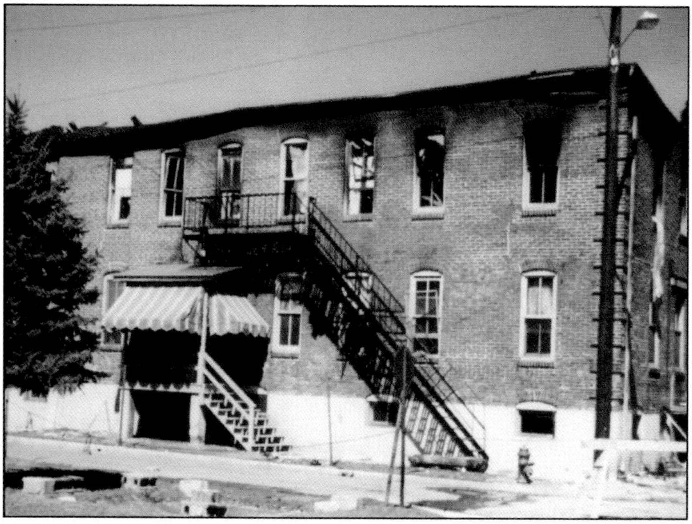

On March 31, 1965, flames broke out in the Faust Apartments at Union Street and Tennis Avenue, south of Parkway in Schuylkill Haven. Nine individuals were left homeless because of the 7:10 p.m. blaze. Two school buses were also destroyed in garages adjacent to the building, which once housed the Sharadin Mill. (R. Nagle, www.schuylkillhavenhistory.com.)

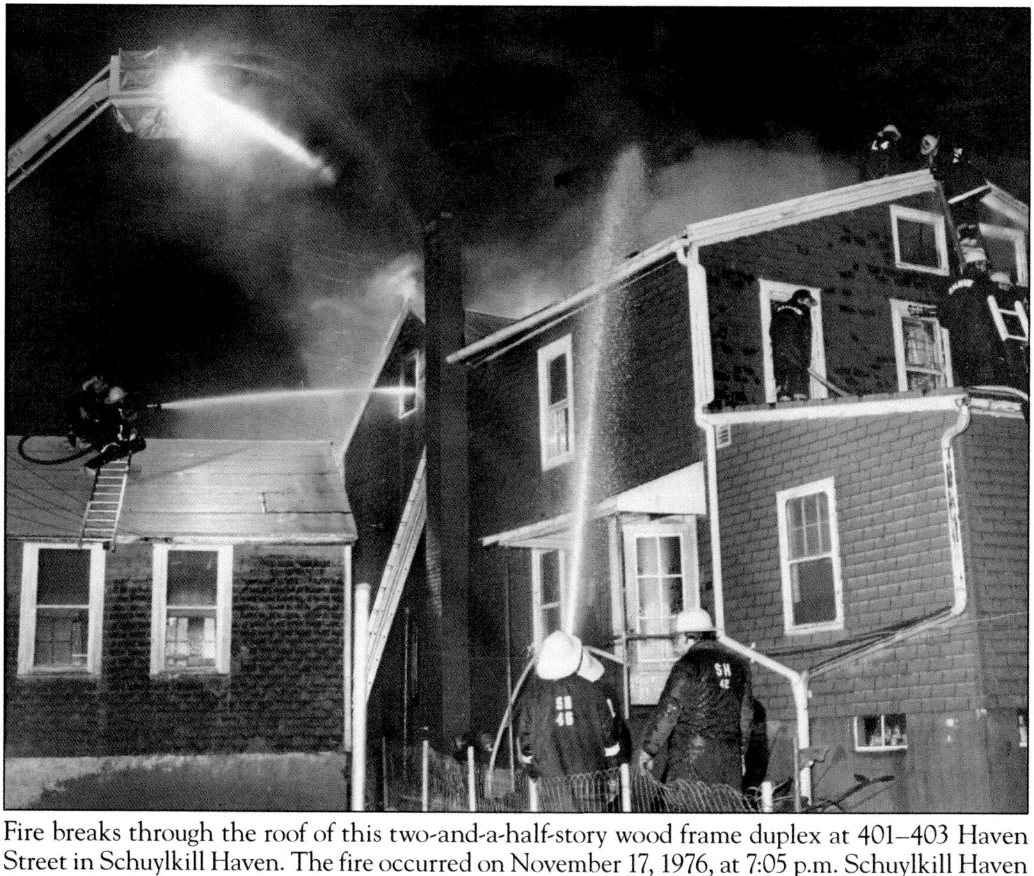

Fire breaks through the roof of this two-and-a-half-story wood frame duplex at 401–403 Haven Street in Schuylkill Haven. The fire occurred on November 17, 1976, at 7:05 p.m. Schuylkill Haven firefighters work over ladders as well as the roof of a nearby garage. The platform of the Rainbow Hose Company snorkel can be seen in the upper left corner of the photograph. (HSSC.)

Heavy smoke rises into the sky as fire destroys the Argo Knitting Mill and Bleachery in Schuylkill Haven on March 7, 1957. The fire in the plant on North Berne Street was discovered at approximately 11:00 a.m. When fire broke through the roof of the building, neighboring homes were threatened by flying brands. (SHFS.)

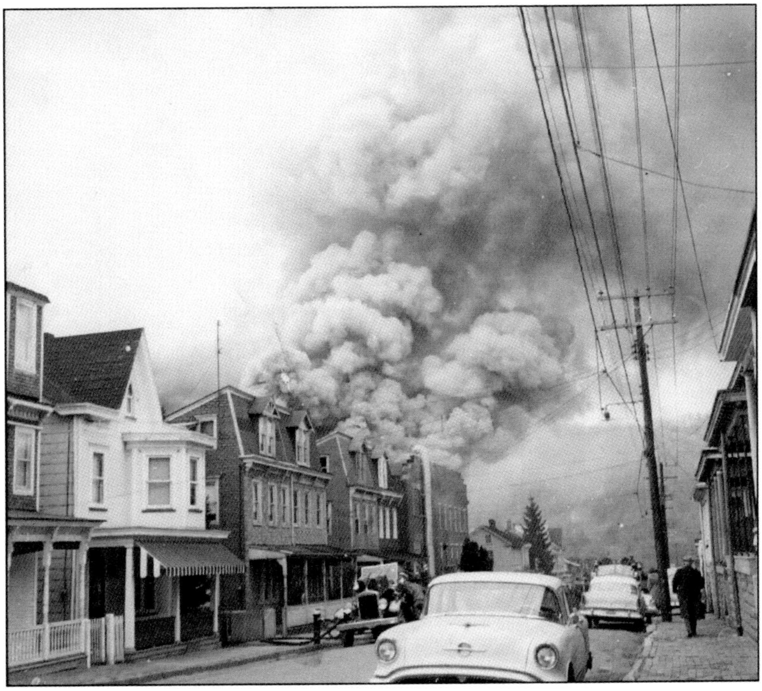

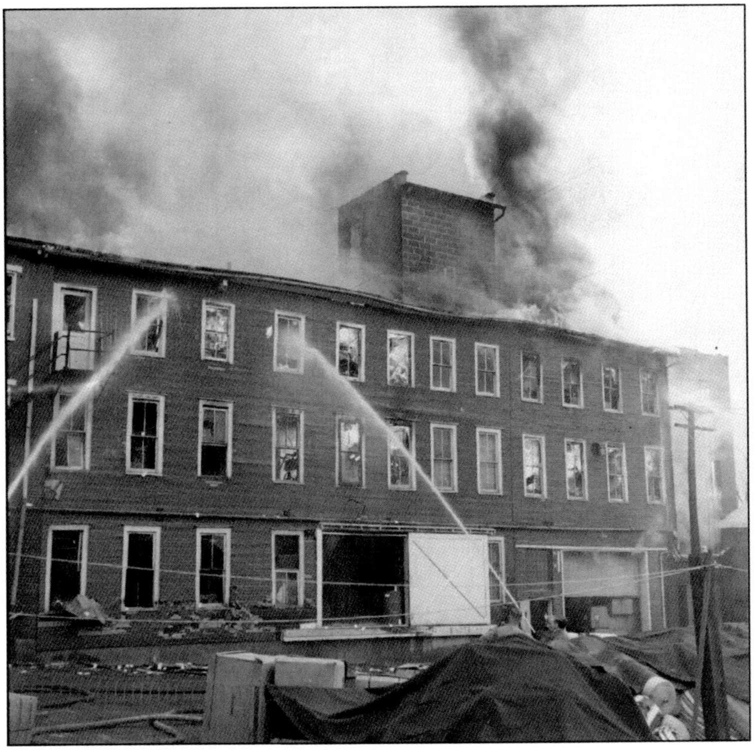

Firefighters direct streams into the heavily involved interior of the Argo Knitting Mill in Schuylkill Haven. The fire was believed to have started above a dryer and spread into the partition wall. This may explain why the building's sprinkler system, while functioning, had little effect on the fire. Fire companies from Pottsville and Orwigsburg assisted the Schuylkill Haven Fire Department. (SHFS.)

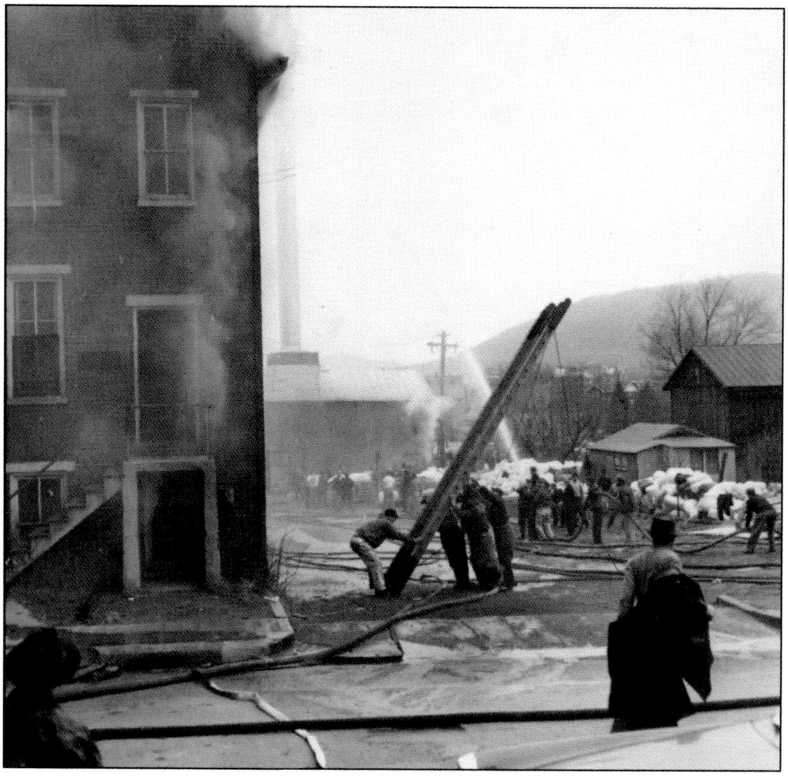

As several firefighters direct a 2.5-inch hose stream into the upper floors of the Argo Mill, others raise an extension ladder. A machinist employed by the mill was briefly trapped on the third floor until he was spotted by a passerby and rescued via ladder by firefighters. One fireman suffered a head injury caused by falling debris. Damage was estimated at $175,000. (SHFS.)

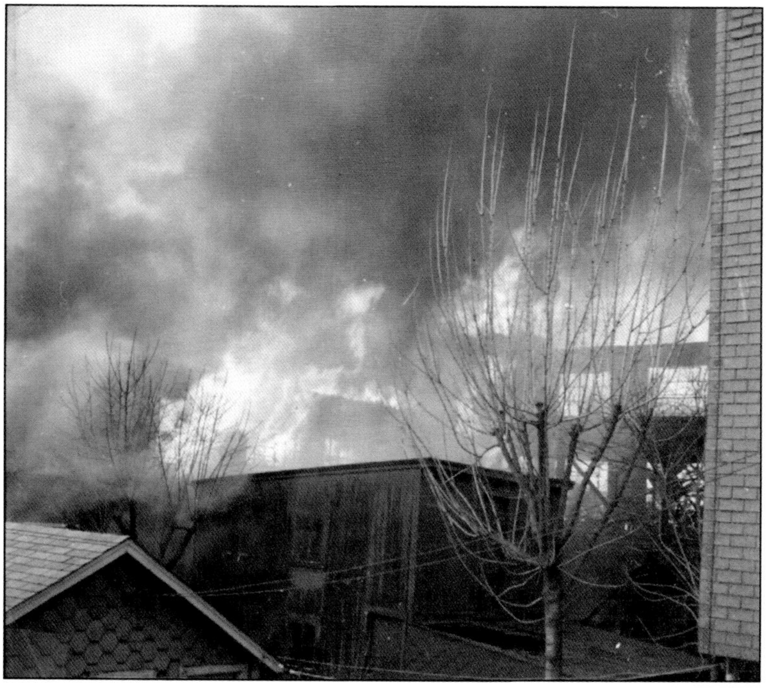

Garages and storage sheds to the rear of the Argo Knitting Mill are threatened by heavy fire blowing from the windows of the plant. Firefighters prevented the fire from seriously damaging the exposed buildings surrounding the plant. Water was supplied from nearby Stoyer's Dam in addition to municipal hydrants. A playground now occupies the site where the mill had stood. (SHFS.)

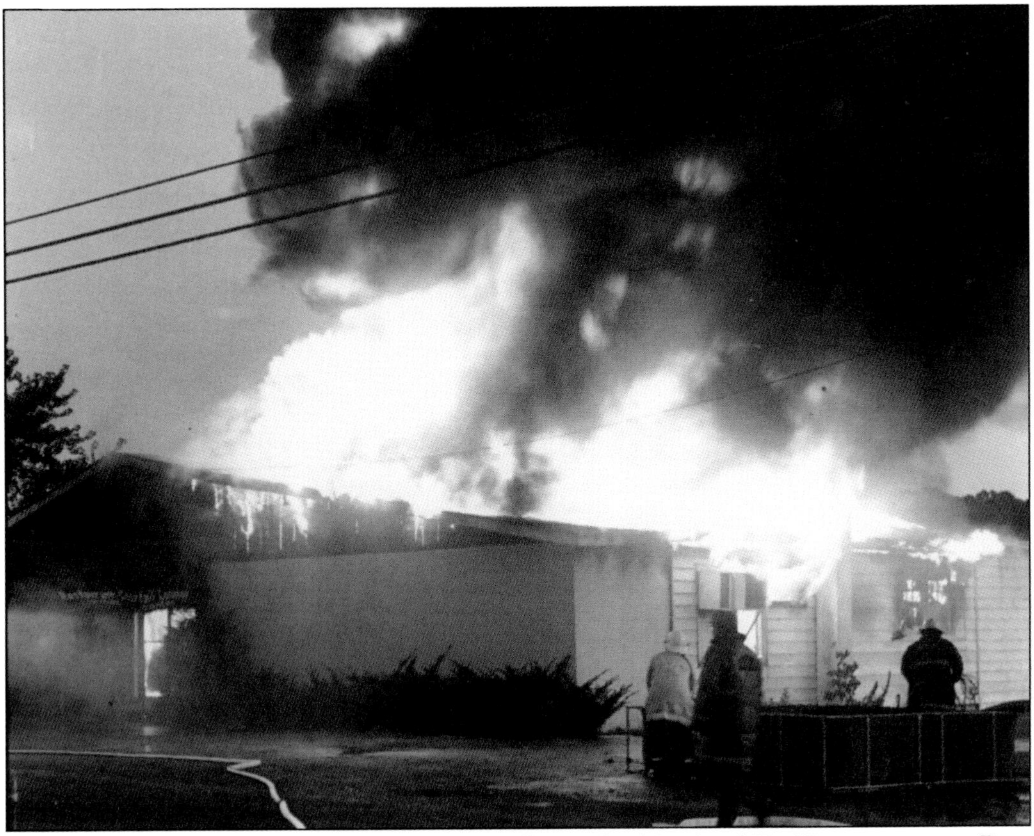

A portable dam is set-up near the front of the fully involved Surf and Turf Restaurant near Deer Lake on September 9, 1975. Twelve pieces of fire apparatus and approximately 60 firefighters fought the blaze. A passing truck driver discovered the fire and stopped at a nearby restaurant, where a waitress telephoned the alarm. (HSSC.)

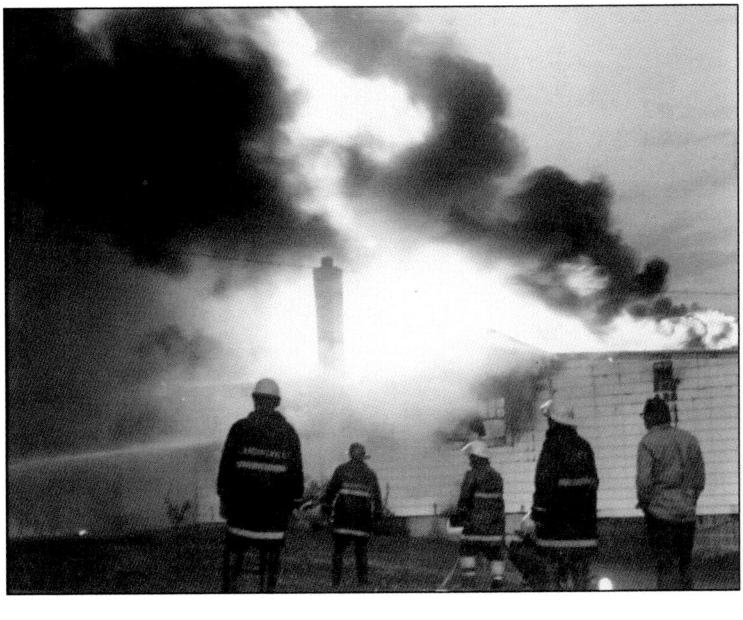

Orwigsburg firefighters arrived to find heavy fire in the basement and kitchen area of the Surf and Turf Restaurant on Route 61 on Tuesday, September 9, 1975. Mutual aid was summoned from Deer Lake, Auburn, Landingville, and Schuylkill Haven. Tankers shuttled water to the scene. (SHFS.)

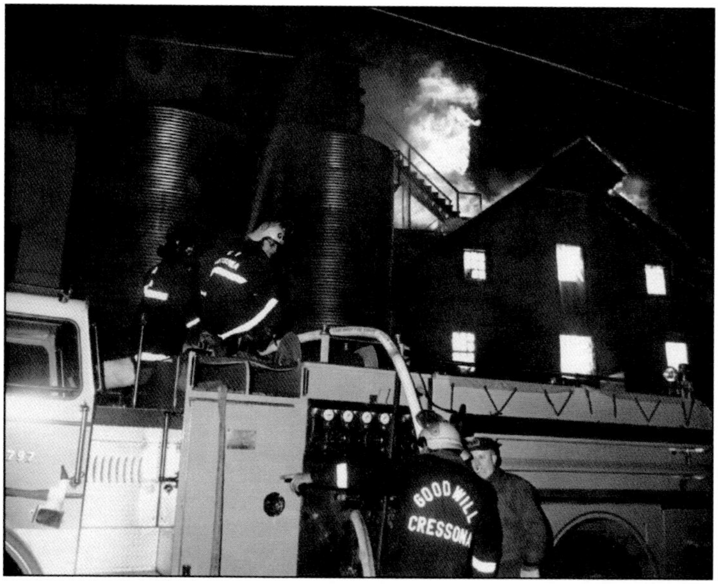

Firefighters from the Good Will Hose Company in Cressona operate the deluge gun atop their 1970 Hahn pumper as fire consumes the Bay-Mor Pet Feed plant on the Gordon Nagle Trail in Cressona. The fire began at approximately 8:00 p.m. on March 17, 1976. Temperatures at the time of the fire were below freezing. (HSSC.)

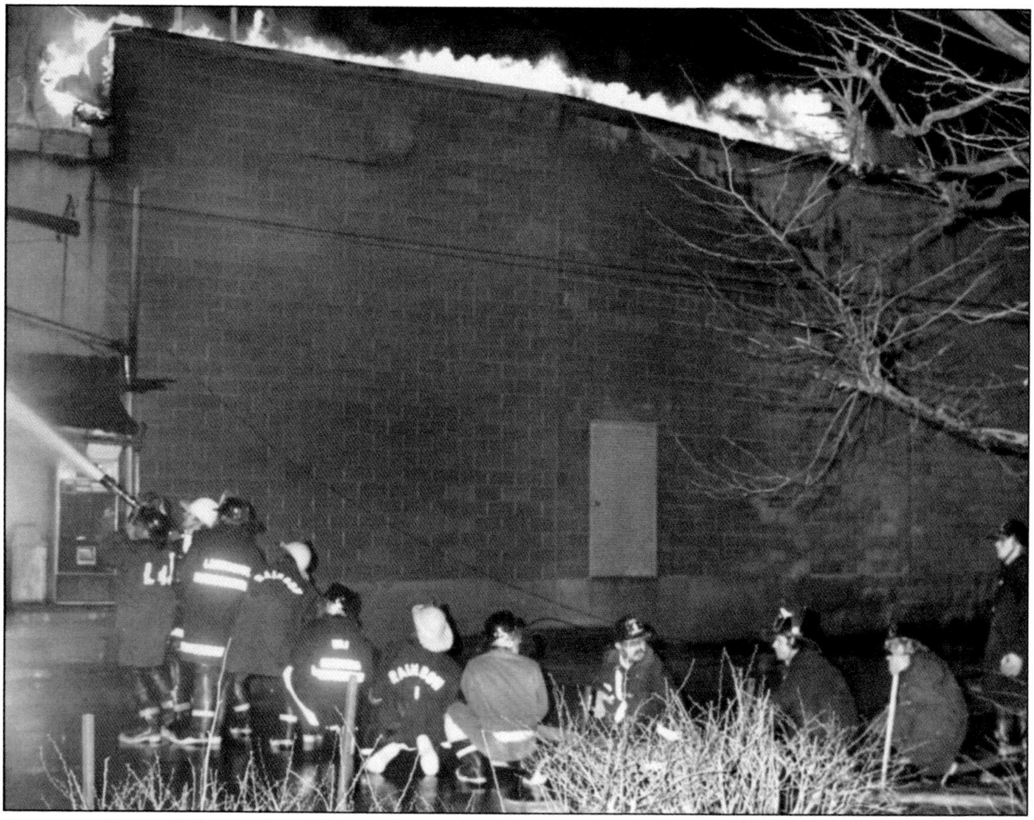

A stream from a 2.5-inch hand line is directed by firefighters as flames destroy much of the Bay-Mor Pet Feed plant in Cressona. In addition to fire department tankers, water was also supplied from the West Branch of the Schuylkill River. Damage to the plant was estimated at $750,000. Assisting Cressona companies were fire companies from Schuylkill Haven, Orwigsburg, Landingville, Pine Grove, Friedensburg, and Summit Station. (HSSC.)

Schuylkill Haven firefighters overhaul the gutted office and parts section of the Schuylkill Brake and Equipment Company. The building, located on Route 61 just south of Schuylkill Haven, was destroyed on December 3, 1973. The alarm was transmitted at 9:50 a.m. A malfunctioning oil burner was suspected as having caused the blaze. All three Schuylkill Haven companies fought the fire. (HSSC.)

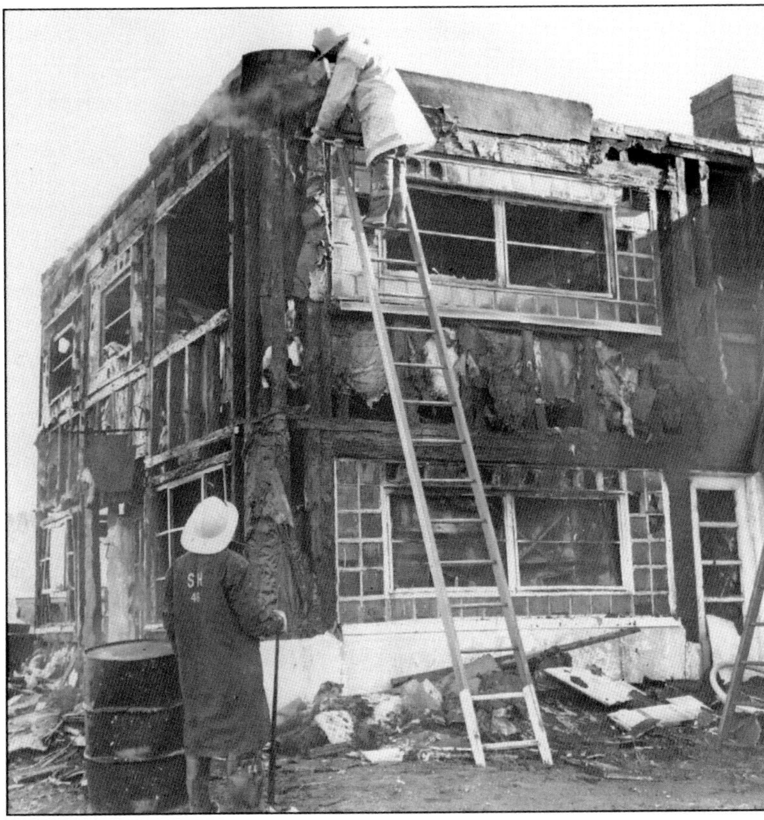

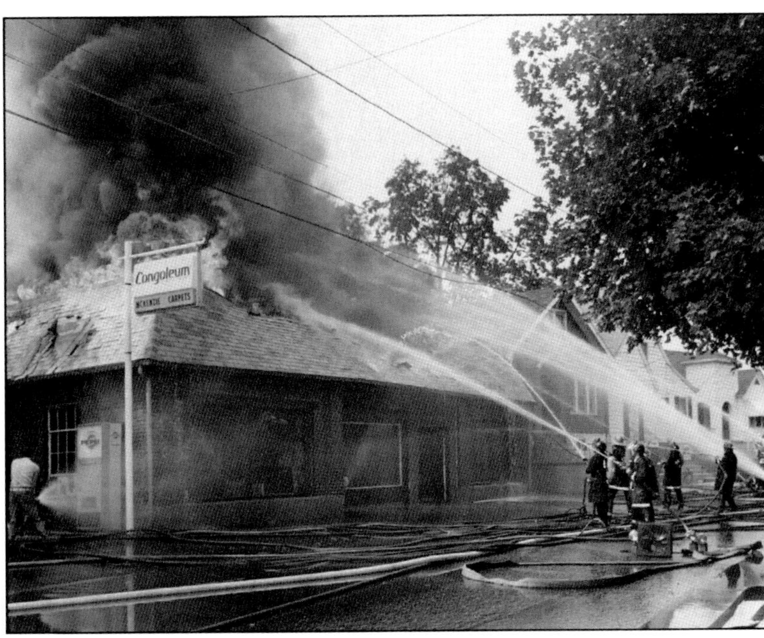

Flames consume the roof of Bob McKenzie's Carpets and Tile business on West Main Street in Schuylkill Haven as firefighters direct numerous streams into the building. The fire began at approximately 3:00 p.m. on August 7, 1974. Firefighters from the three Schuylkill Haven fire companies were on the scene until 6:00 p.m. The business had opened only eight months before the fire. (HSSC.)

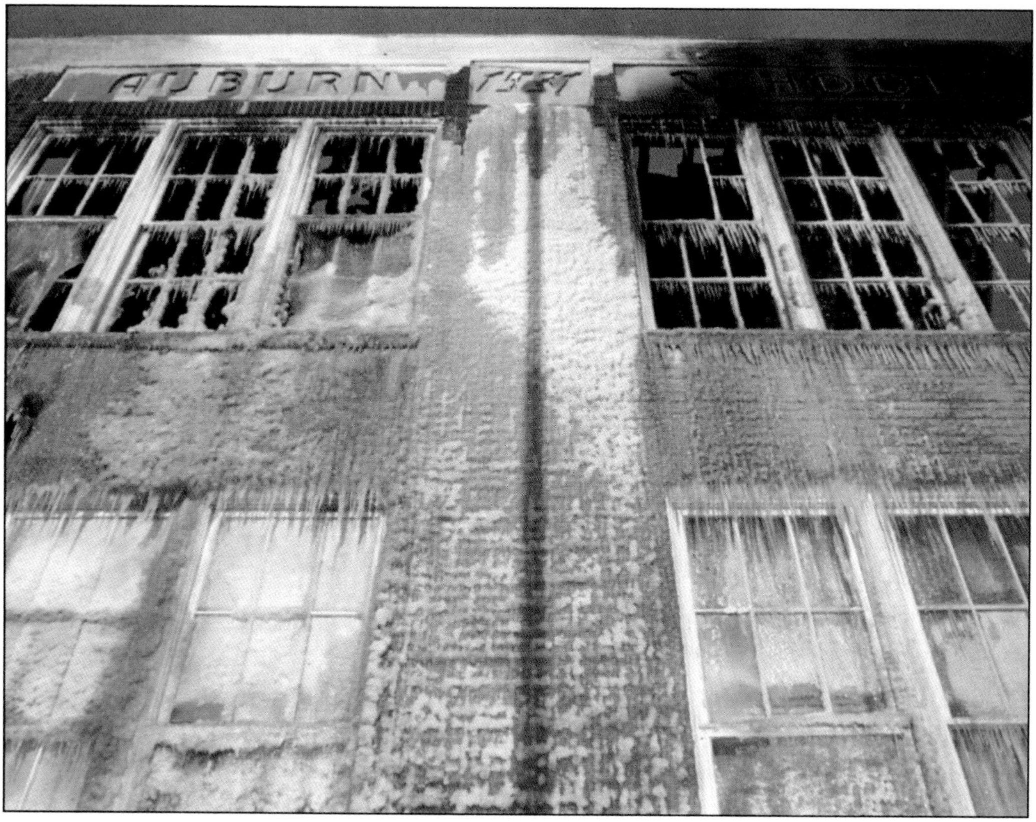

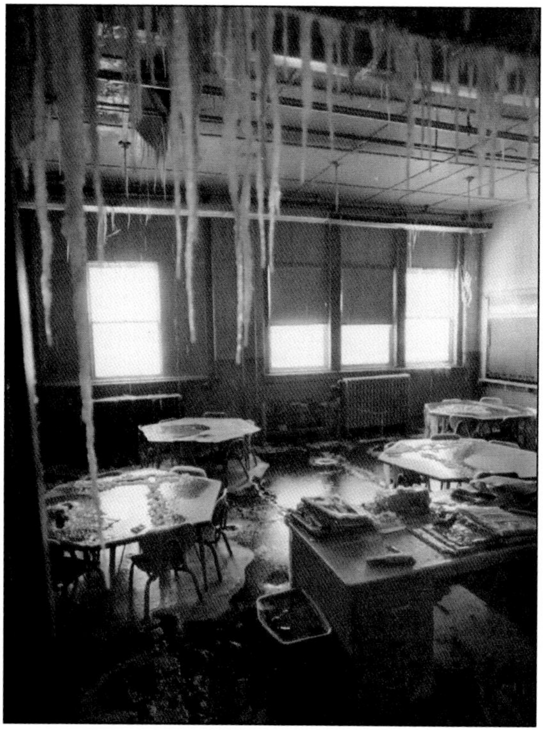

On the day after Christmas in 1983, firefighters from Auburn and throughout southern Schuylkill County were challenged when fire struck the Auburn Elementary School. Freezing temperatures resulted in a large accumulation of ice on the building and on the scene. The wind chill at the time of the 12:45 a.m. alarm was -20 degrees. (HSSC.)

Ice coats everything in this first-floor classroom at the Auburn Elementary School after a December 26, 1983, fire. The school housed 88 students, three full-time teachers, and one part-time teacher at the time. Thankfully it was closed for Christmas break when the fire occurred. Fire companies from Auburn, Landingville, Deer Lake, Schuylkill Haven, and Orwigsburg responded. (HSSC.)

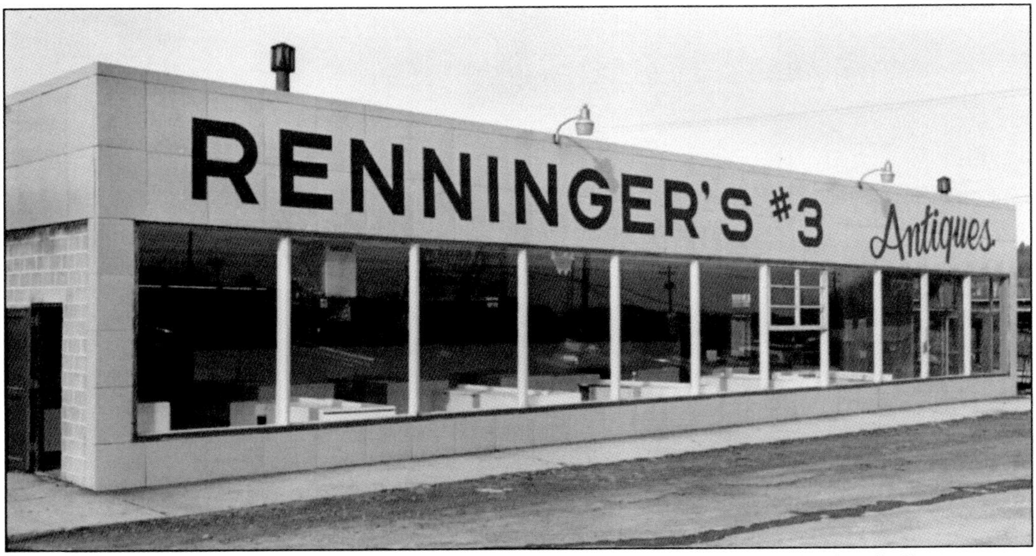

Renninger's Farmers Market was a local landmark situated just south of Schuylkill Haven on Route 61. While the Renninger's Three Antique Store was of block construction, most of the one-story market was wood frame. The market was built in 1942 and was approximately 300 feet long. (HSSC.)

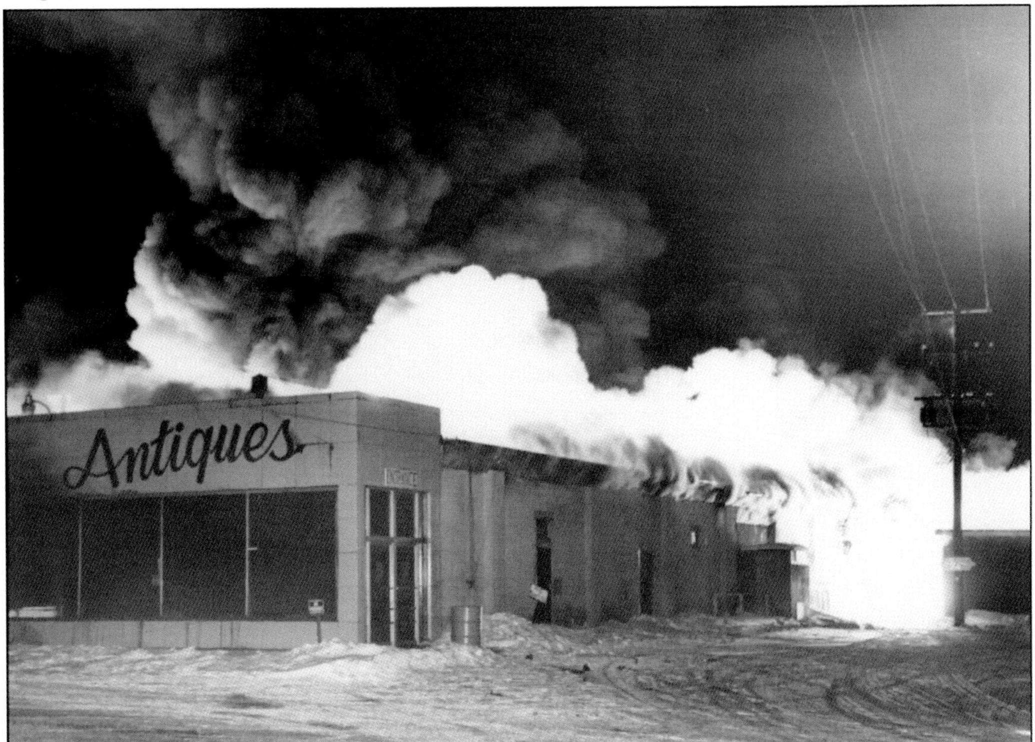

On Friday, January 22, 1982, fire erupted at Renninger's Farmers Market. The Schuylkill Haven Fire Department received the initial call at 6:48 p.m. On arrival, Chief Kenneth "Butch" Reed was confronted with a well-advanced fire in the frame building, extending quickly into the block antique store. Heavy fire conditions are evidenced in the photograph. (HSSC.)

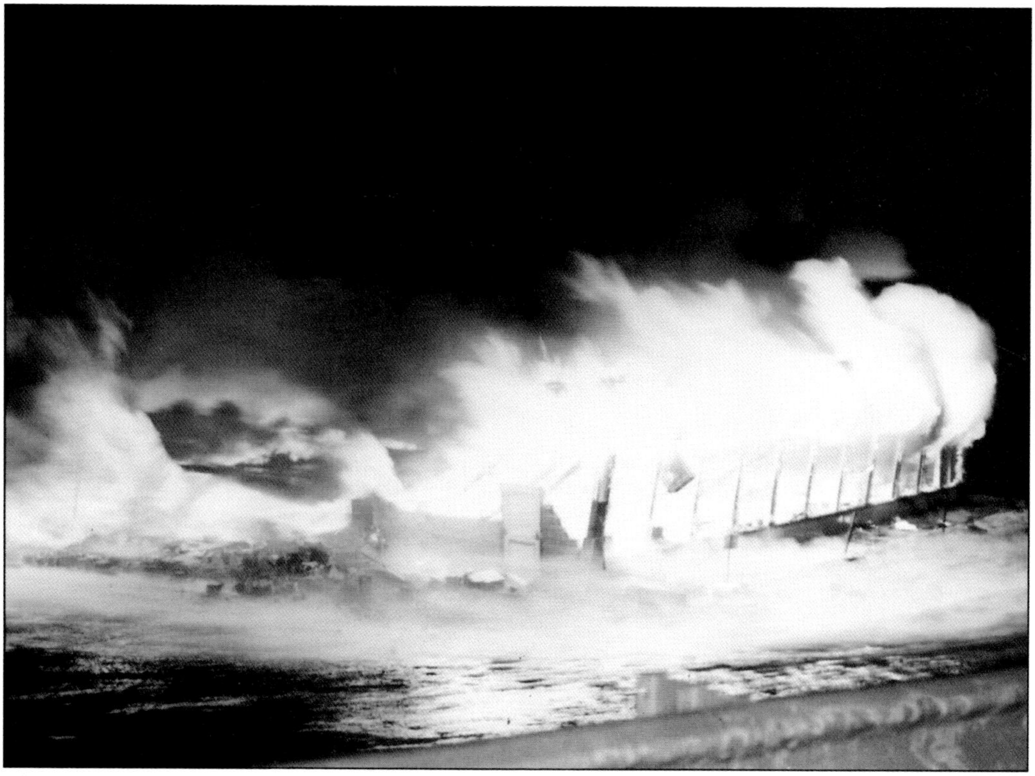

The Renninger's Three portion of the Farmers Market is reduced to rubble during the January 22, 1982, fire. A lack of fire hydrants in the area necessitated firefighters using fire department tankers to shuttle water to the scene. Water was obtained from a dam owned by Renninger's, located approximately a half mile away, as well as a stream located near the Schuylkill Country Club golf course. (HSSC.)

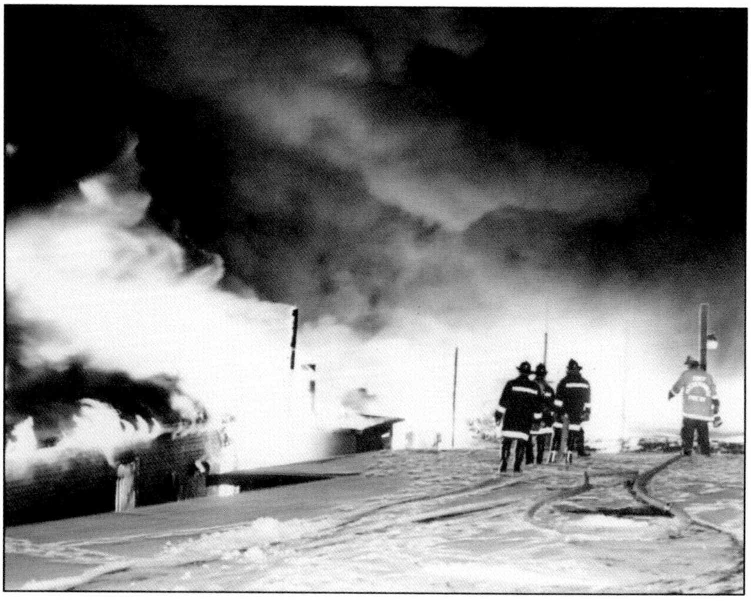

Firefighters from Friedensburg wait for water on the roof of the outdoor market place, an open-air pavilion, at Renninger's Farmers Market as fire destroys the frame market building proper. In addition to the three Schuylkill Haven fire companies, firefighters from Deer Lake, Orwigsburg, New Ringgold, Cressona, Landingville, Summit Station, and Friedensburg assisted on scene. (HSSC.)

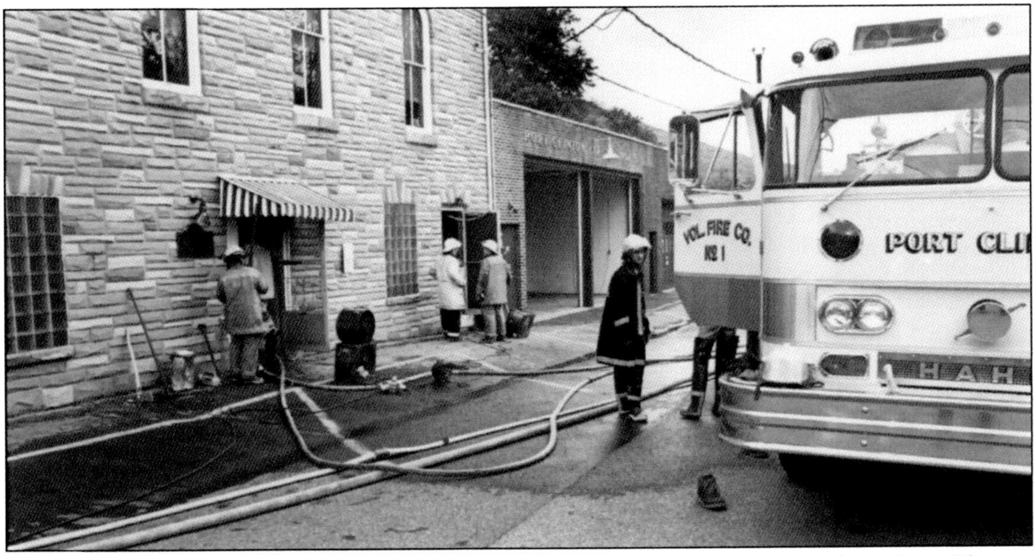

This is a little too close for comfort! On June 4, 1990, the social quarters of the Port Clinton Volunteer Fire Company No. 1 were damaged by fire. The engine room, where the apparatus was housed, was not damaged. Port Clinton's Hahn pumper had a short run to get in service for this fire. Hamburg and Deer Lake assisted Port Clinton firefighters. (HSSC.)

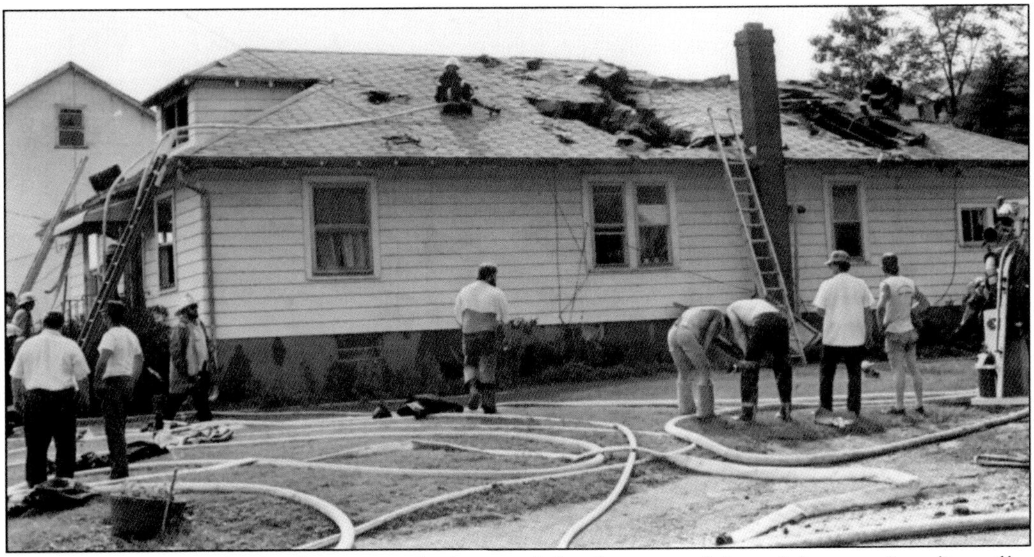

Fire originating in the kitchen severely damaged this home in Adamsdale, near Landingville. The blaze occurred on June 14, 1983, at approximately 3:00 p.m. With the fire controlled, several firefighters take a break on a warm June afternoon as others continue with the overhaul tasks. Eight fire companies assisted in fighting the blaze. (HSSC.)

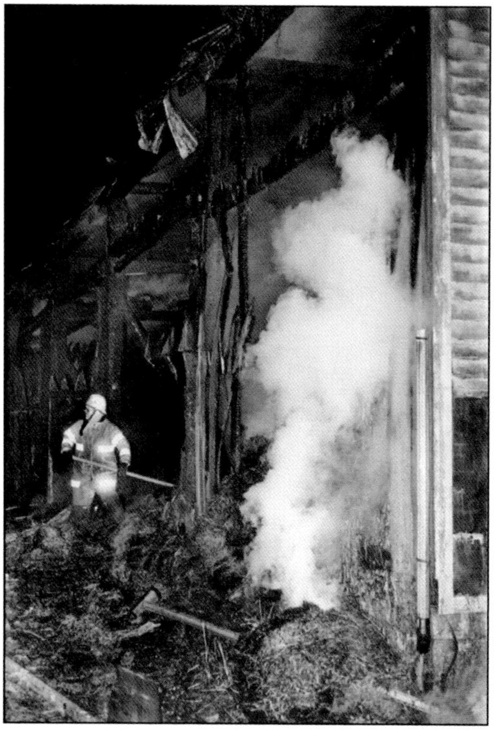

On Saturday, April 10, 1982, at 10:25 p.m., the alarm of fire was sounded for the Heim barn, located in McKeansburg between Routes 895 and 443. The barn was fully involved when firefighters arrived. Seven fire companies assisted as tankers were used to transport water to the scene. Firefighters were on the scene until 5:00 a.m. April 11, Easter Sunday. (HSSC.)

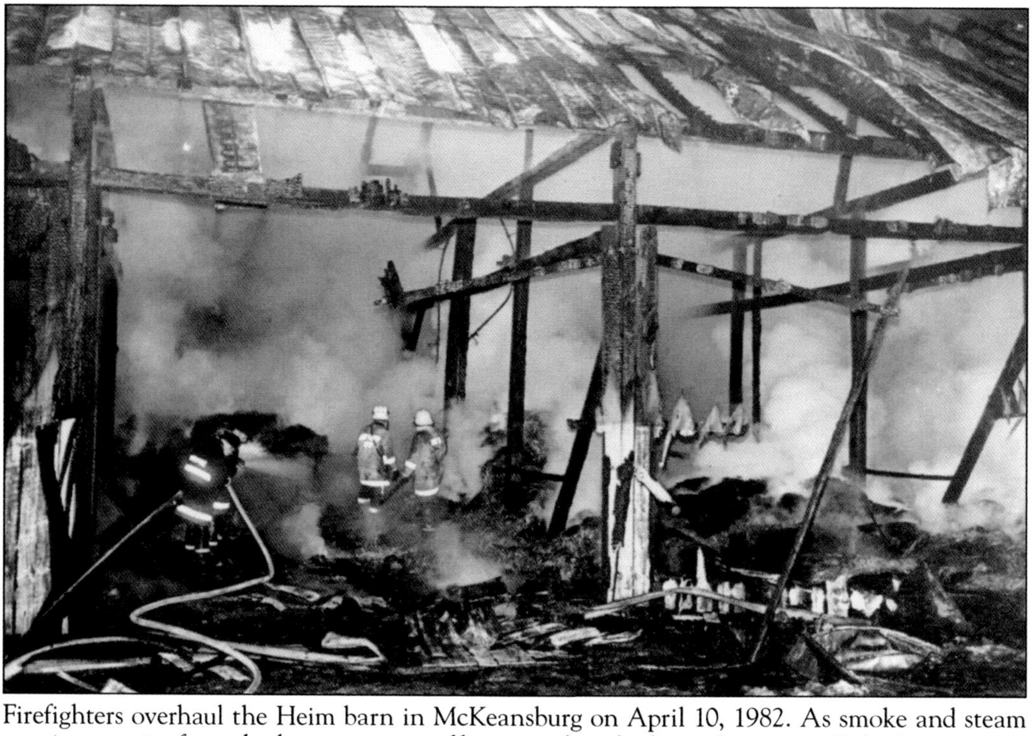

Firefighters overhaul the Heim barn in McKeansburg on April 10, 1982. As smoke and steam continue to rise from the large quantity of hay stored in the barn, firemen pull the hay apart to expose burning pockets and completely soak them. With the visible severe damage to the barn structure, fire conditions when the first unit arrived must have been extreme. (HSSC.)

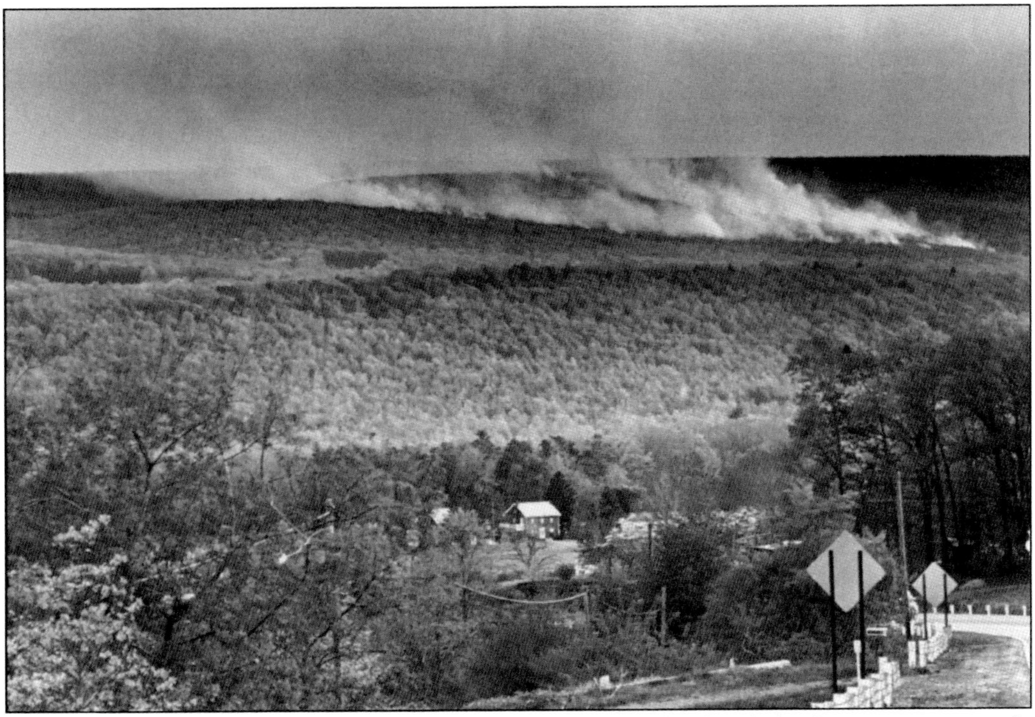

Heavy smoke over the treetops is evident from Route 183 as this brush fire burns approximately 165 acres of woodland—most of it owned by the Hawk Mountain Council of the Boy Scouts of America. The fire began approximately 1:30 p.m. on May 11, 1982. More than 100 firefighters were on the scene from nine companies in both Berks and Schuylkill Counties. (HSSC.)

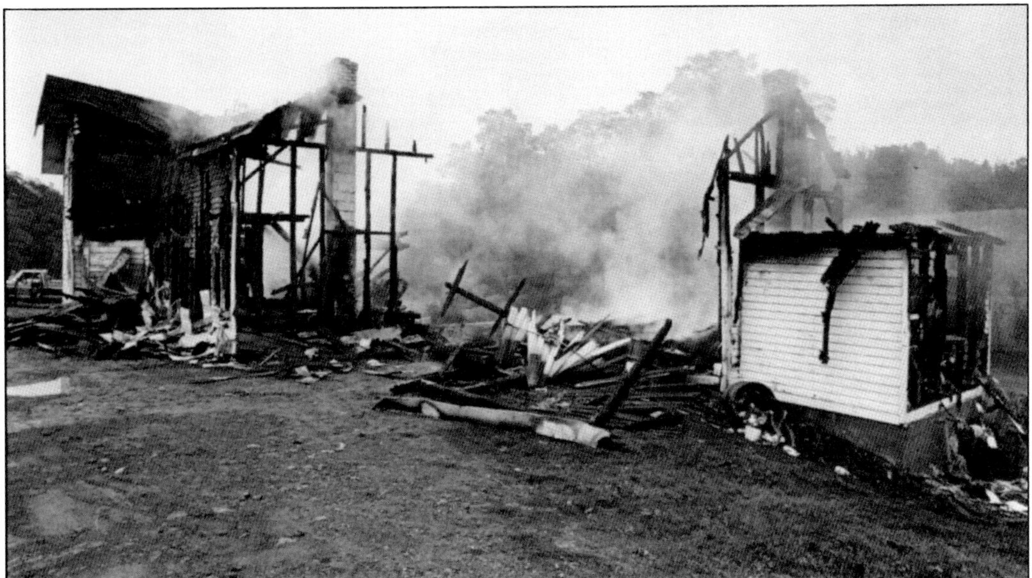

A family of eight was left homeless on September 20, 1982, when fire completely destroyed their two-story frame home on Stanhope Road in Pine Grove Township. Damage was estimated at $40,000. Firefighters from Ravine, Pine Grove, Friedensburg, and Summit Station fought the 10:37 a.m. fire. (HSSC.)

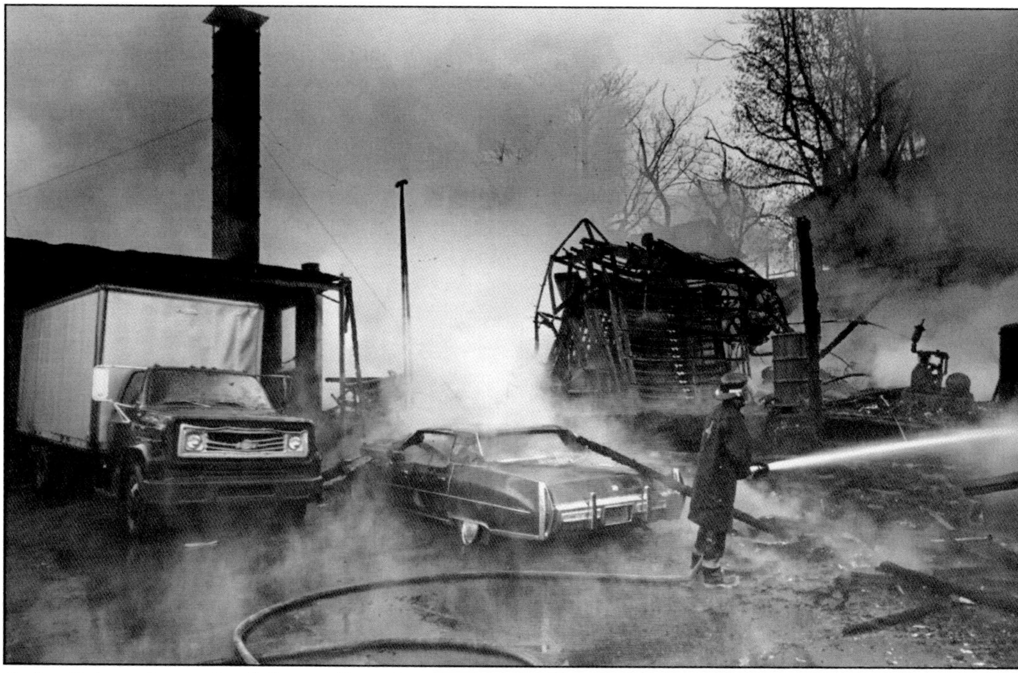

A firefighter from the Liberty Fire Company of Schuylkill Haven wets down the ruins of the Schuylkill Haven Bleach and Dye Company on May 1, 1982. The fire was believed to have been caused by a dust or lint explosion near a large drying machine. At the height of the fire, flames were shooting 40 to 50 feet above the single-story building. (HSSC.)

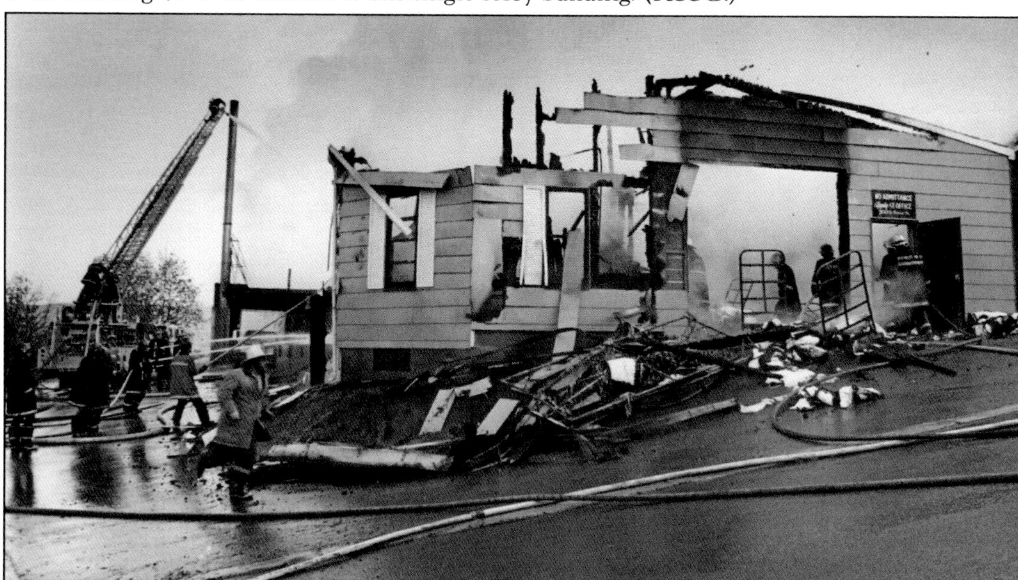

West End Hose Company of Pottsville has the ladder pipe in service on its 1977 Seagrave aerial ladder truck as crews continue to overhaul the Schuylkill Haven Bleach and Dye Company. The alarm was sounded at 5:00 a.m. on May 1, 1982. The building was located at Market and St. Peter Streets in the borough. Firefighters from Orwigsburg and Pottsville assisted Schuylkill Haven firefighters on the scene. (HSSC.)

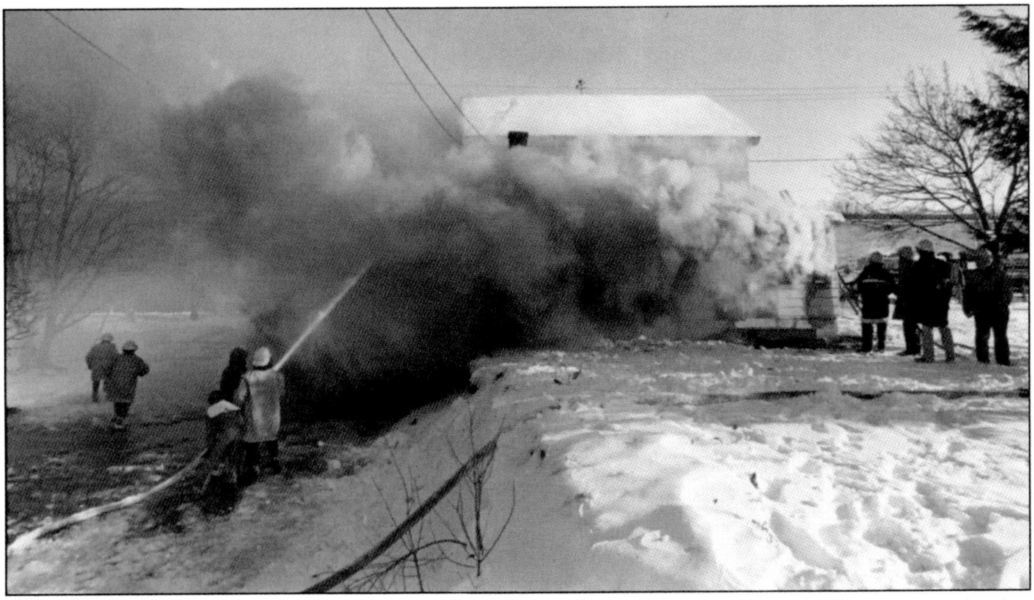

Heavy smoke pours from this two-story frame house on Route 895 near Summit Station. The fire occurred at approximately 4:00 p.m. on April 7, 1982, and began in the basement garage. One occupant of the home suffered burn injuries. Six fire companies battled the blaze and the early April snow accumulation. (HSSC.)

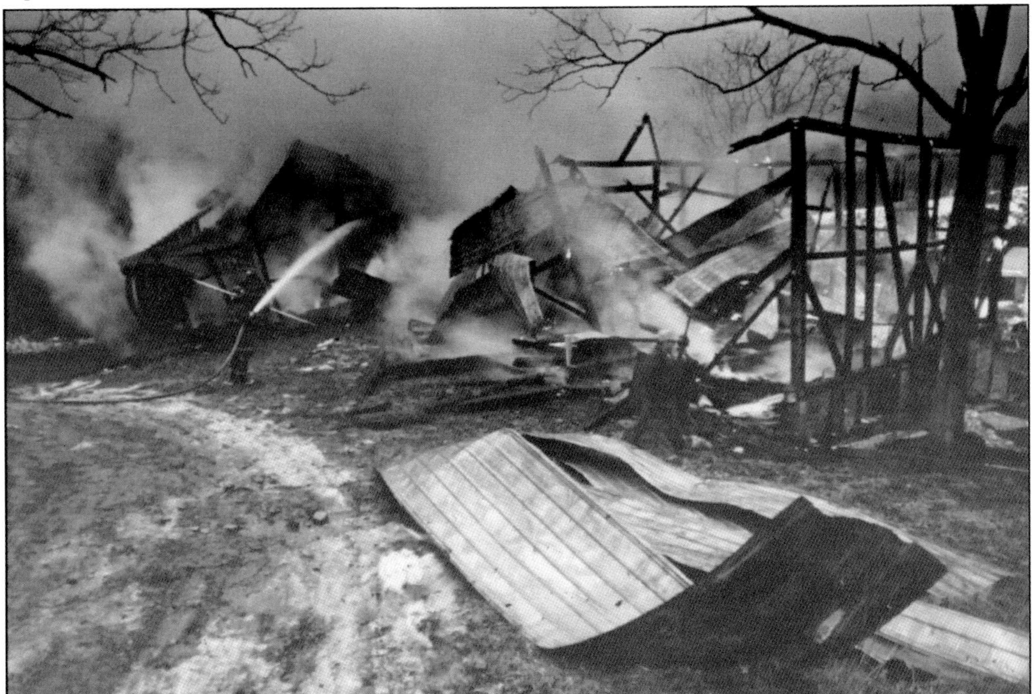

Fire destroyed this 200-year-old, hay-filled barn on Stanhope Road in Pine Grove Township on February 3, 1984. The wooden support beams in the barn were all hand cut. The blaze, which occurred at approximately 4:00 p.m., was ruled as an arson. The total loss was placed at over $100,000. (HSSC.)

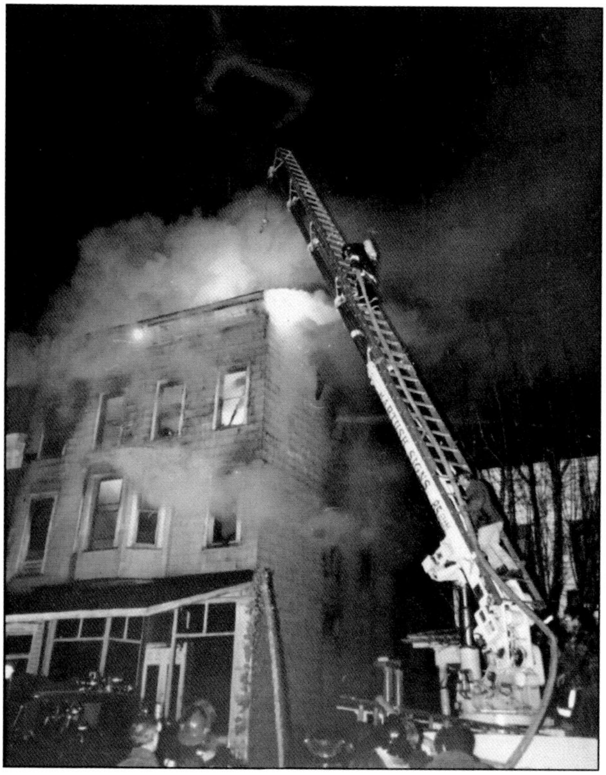

Improvise, adapt, and overcome: a firefighter stretches a hose line over an 85-foot crane owned by the Bartush Sign Company of Orwigsburg at this fire that began at 202 East Market Street in Orwigsburg. Firefighters pressed the Bartush crane into service prior to the arrival of the West End ladder truck from Pottsville. The fire occurred on December 22, 1974, at 2:00 a.m. (HSSC.)

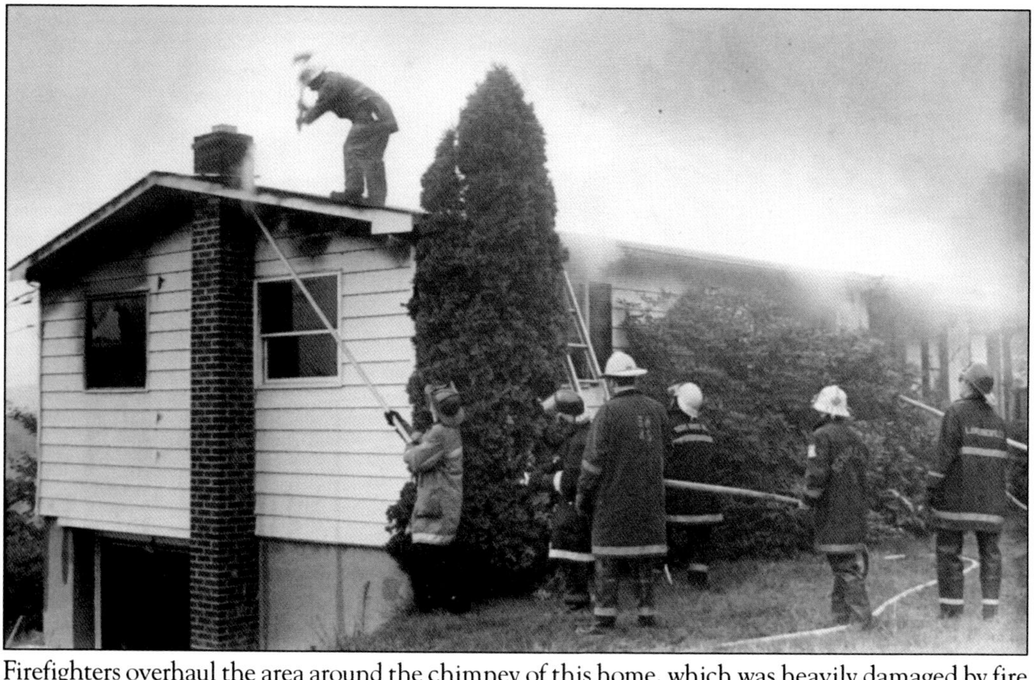

Firefighters overhaul the area around the chimney of this home, which was heavily damaged by fire on August 8, 1982. The house was located in the Renninger's Haven portion of North Manheim Township, just off Route 61. All three Schuylkill Haven companies, as well as Landingville's Community Fire Company, battled the 10:30 a.m. fire. (HSSC.)

Faced with a severe water shortage in November 1980, the Schuylkill Haven Fire Department devised a plan to use tankers to transport water from Stoyer's Dam in the borough to a fire scene. At 8:50 a.m. on November 19, 1980, the plan was tested when fire destroyed the First United Methodist Church on St. John Street in the borough. Rainbow Hose Company's snorkel is positioned in the photograph. (HSSC.)

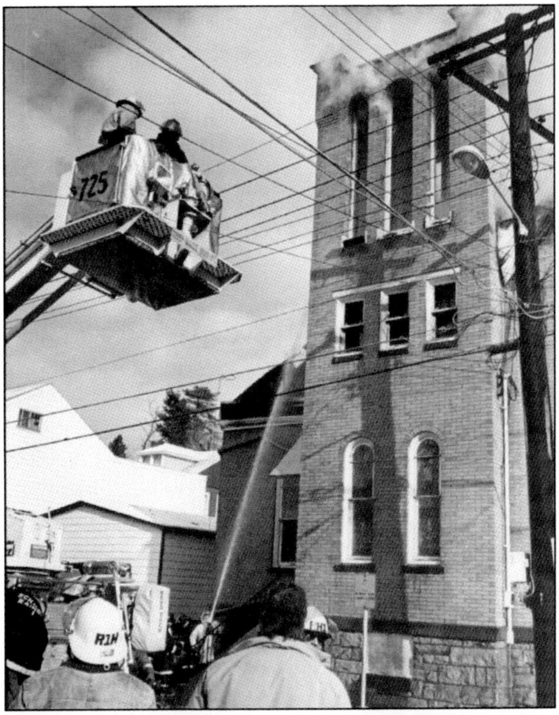

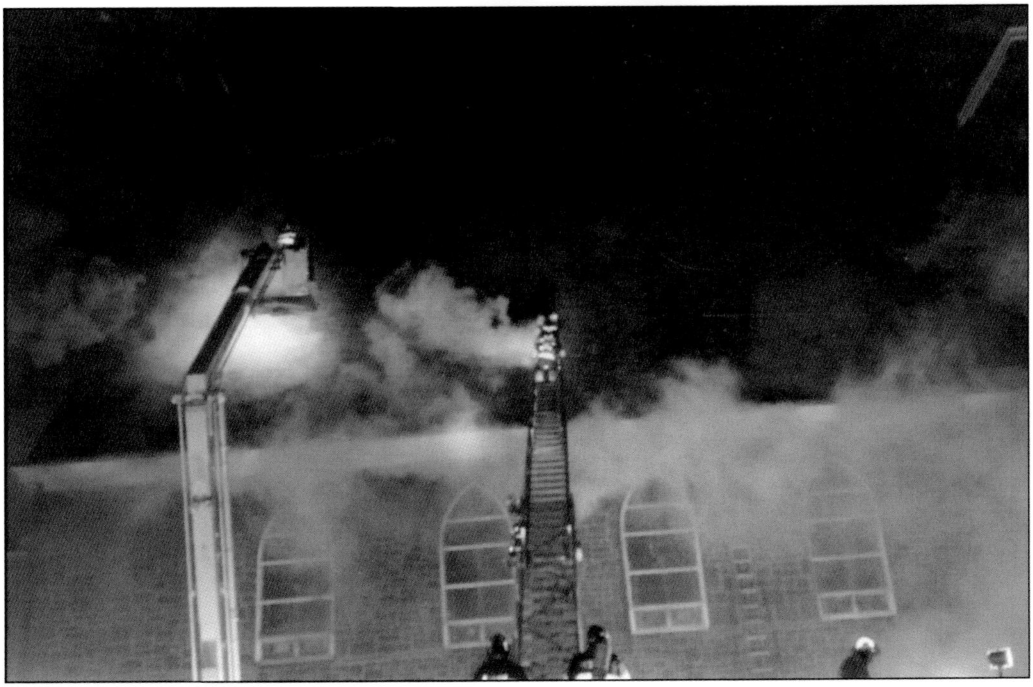

Built in 1905, the Christ Evangelical Lutheran Church at 252 Dock Street in Schuylkill Haven was destroyed by fire on Sunday, December 29, 1991. Heavy smoke pushes from the eaves of the church as firefighters work to ventilate the roof from the Rainbow Hose Company's 55-foot snorkel (left) and the Rescue Hook and Ladder Company of St. Clair's 65-foot aerial ladder (right). The alarm sounded at 11:05 p.m. (Authors' collection.)

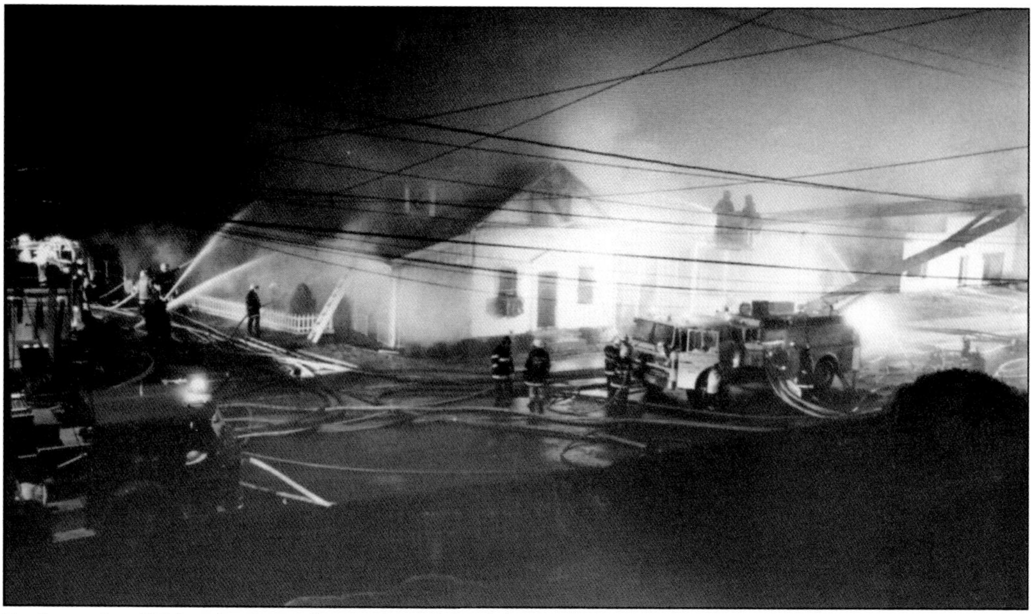

Built in the 1880s, the large frame Canal Building was located at 201 Dock Street in Schuylkill Haven. At 8:00 p.m. on February 29, 1984, the alarm was sounded for a fire that began in the basement and quickly spread throughout. The photograph shows Schuylkill Haven firefighters directing streams from numerous hose lines and the Rainbow Hose Company's snorkel into the building. (HSSC.)

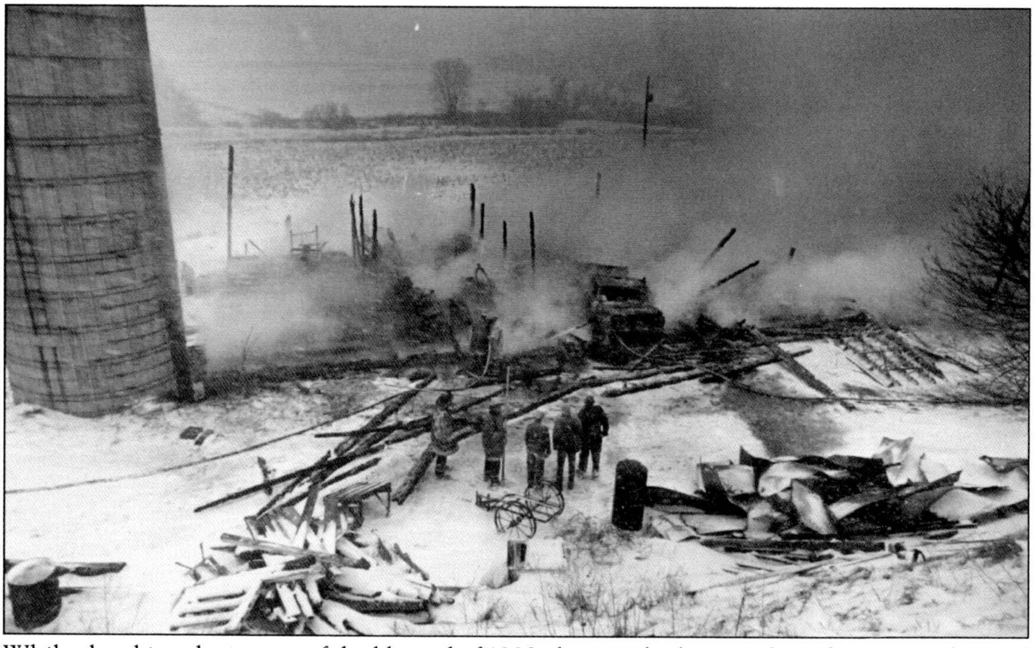

While absorbing the impact of the blizzard of 1983, during which 1 to 2 feet of snow was dumped on Schuylkill County, firefighters from Pine Grove, Friedensburg, Summit Station, Ravine, and Liberty of Schuylkill Haven battled this barn fire in Washington Township. The temperature was 15 degrees when the alarm sounded at 2:30 p.m. on February 11, 1983. Thirty pigs perished in the fire. (HSSC.)

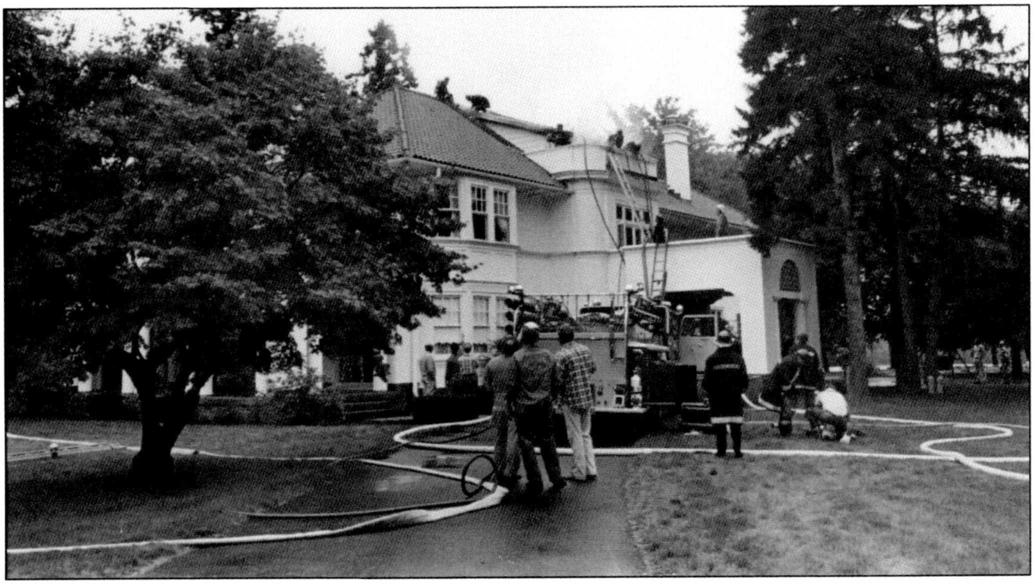

This mansion, built in 1913, was located at Legion Acres near Pine Grove. Fire originating in a recessed ceiling light fixture in the attic hallway caused a fire that heavily damaged the third-floor attic area of the home. The fire occurred on October 1, 1983, at 12:30 p.m. The 1967 International/Hahn pumper of the Hose, Hook, and Ladder Fire Company of Pine Grove is in service. (HSSC.)

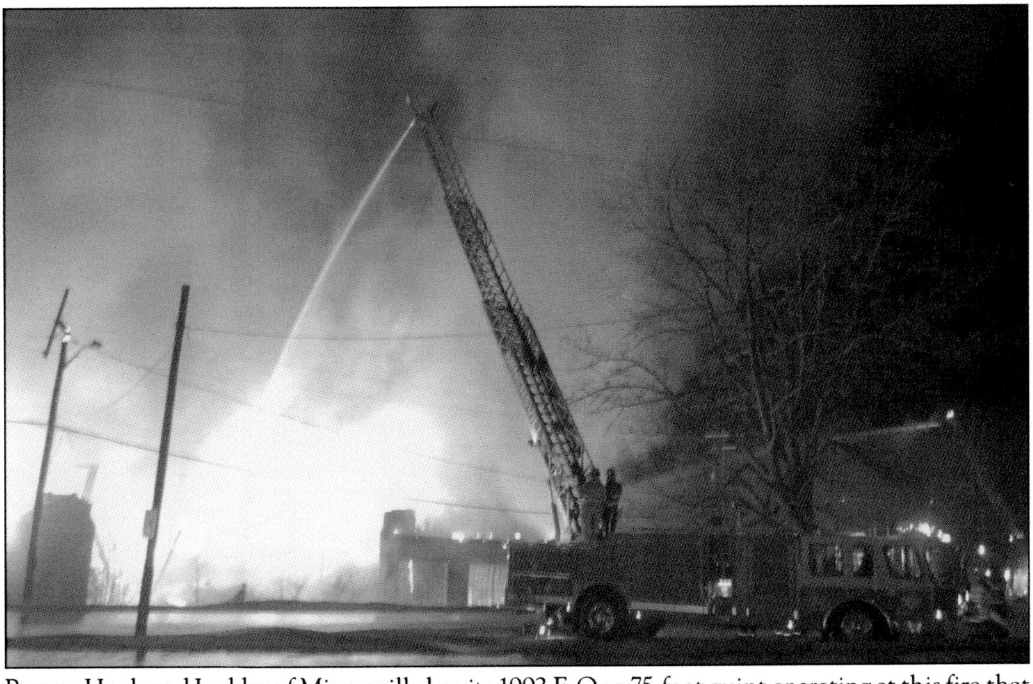

Rescue Hook and Ladder of Minersville has its 1992 E-One 75-foot quint operating at this fire that destroyed the Cressona Textile Waste facility in Cressona on March 26, 2006. The spectacular fire illuminated the sky as it spread rapidly throughout the 300-foot-by-150-foot textile waste recycling plant. Numerous companies from Schuylkill and Berks Counties assisted the Cressona Fire Department. (Dave Hardinger.)

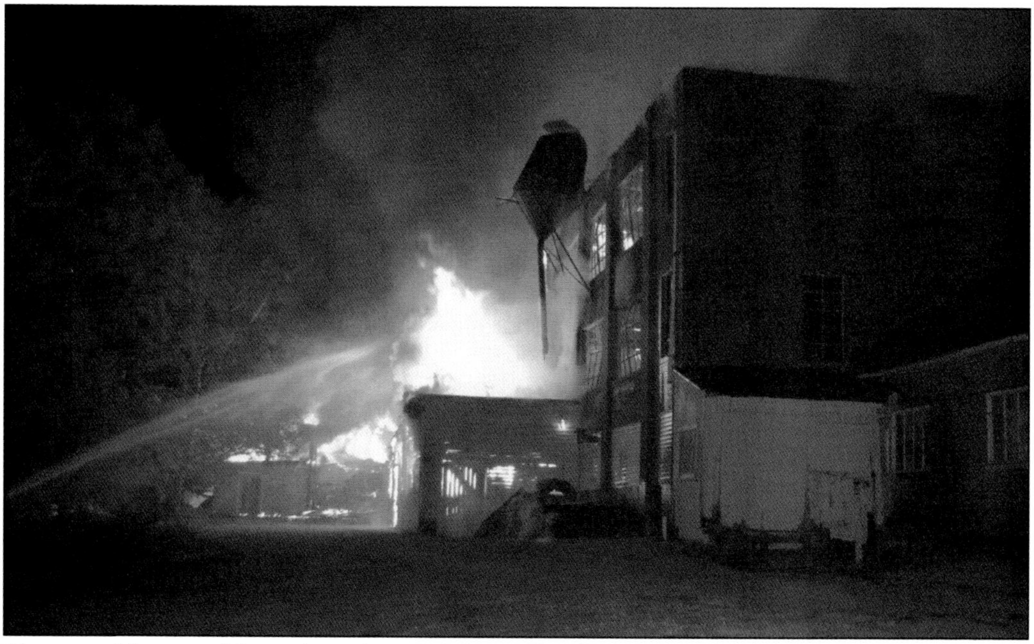

The rear of the very old, very large former Auburn Shoe Factory at 427 Bear Creek Street in Auburn begins to collapse as fire destroys the building on July 17, 2006. The alarm was transmitted at 2:48 a.m. The fire also destroyed an occupied apartment building. Firefighters from throughout southern Schuylkill County as well as northern Berks County assisted Auburn. (Dave Hardinger.)

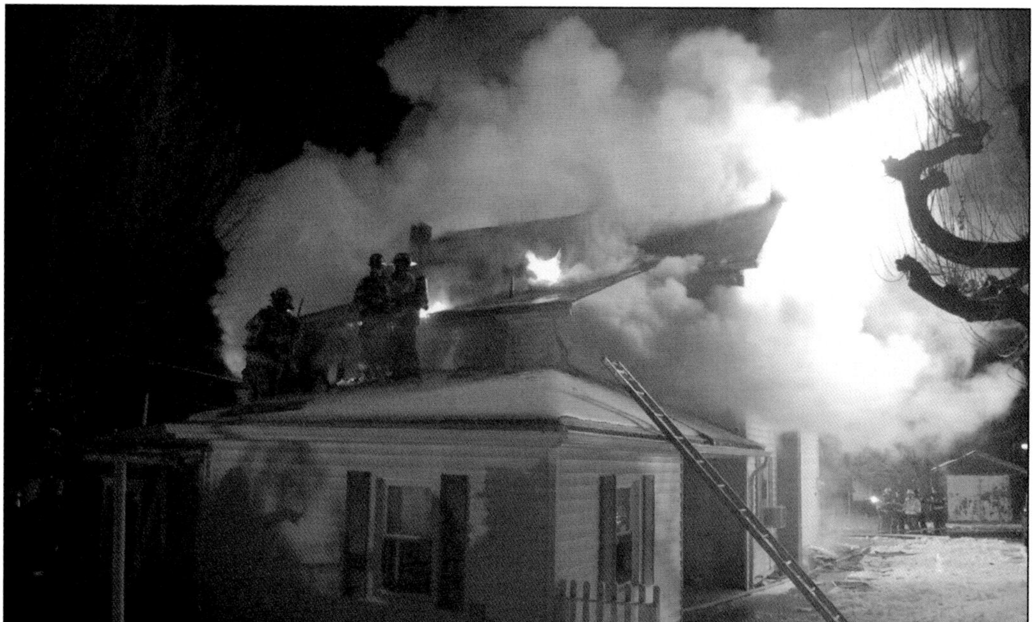

Firefighters work on the snow-covered rear first-floor roof of this two-and-a-half-story frame duplex in Cressona. The fire occurred at approximately 3:00 a.m. on January 31, 2009. Temperatures were hovering around 14 degrees Fahrenheit at the time. One firefighter was injured fighting the fire, which gutted the building. (Dave Hardinger.)

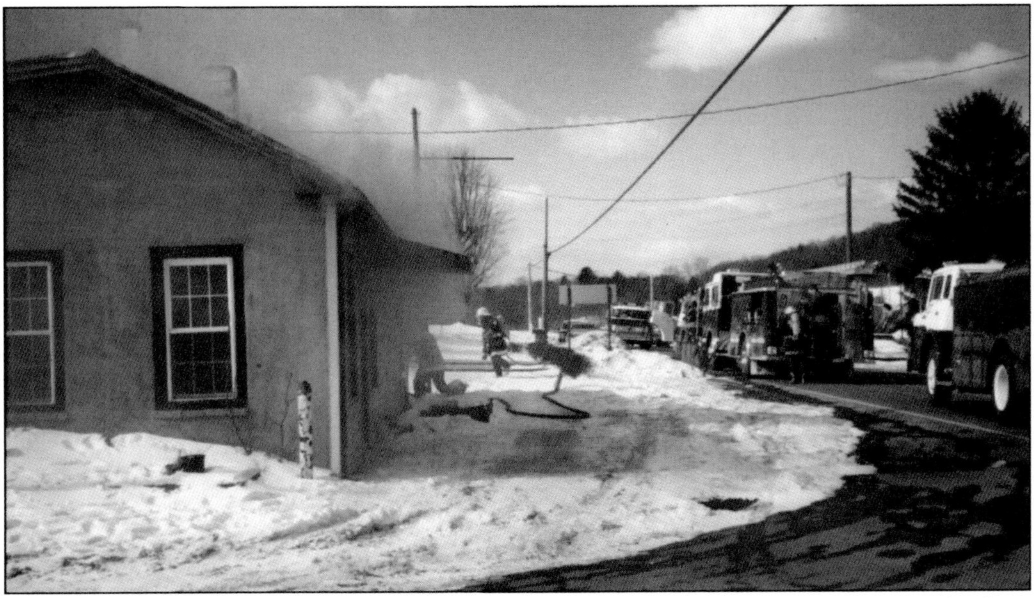

Fire damaged this home at the intersection of Route 443 and Deturksville Road between Friedensburg and Pine Grove during the winter of 1992–1993. Firefighters from Friedensburg are stretching hose lines and setting up their water supply as smoke pours from the home. Crews from Pine Grove and Summit Station also fought the blaze. (Authors' collection.)

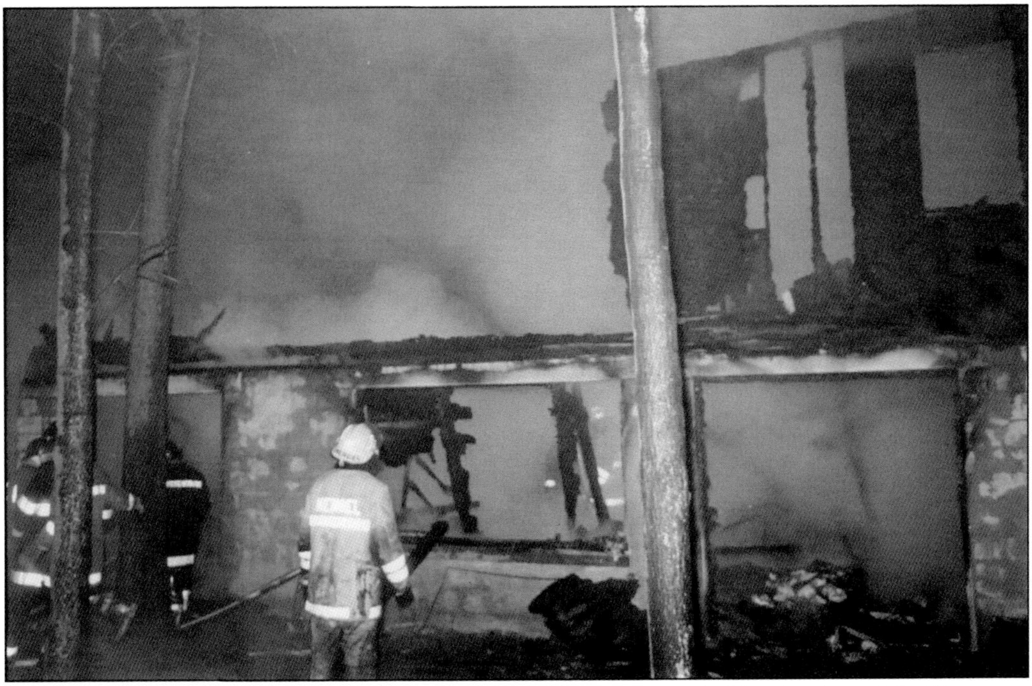

Shortly before 11:00 p.m. on February 11, 1984, the alarm was sounded for this fire at 770 Cheyenne Drive in Lake Wynonah. When firefighters from Friedensburg, Summit Station, Landingville, and Liberty of Schuylkill Haven arrived, fire had already burned through the roof of the home. As evidenced by the photograph, the home was completely destroyed. The occupants were not home at the time. (HSSC.)

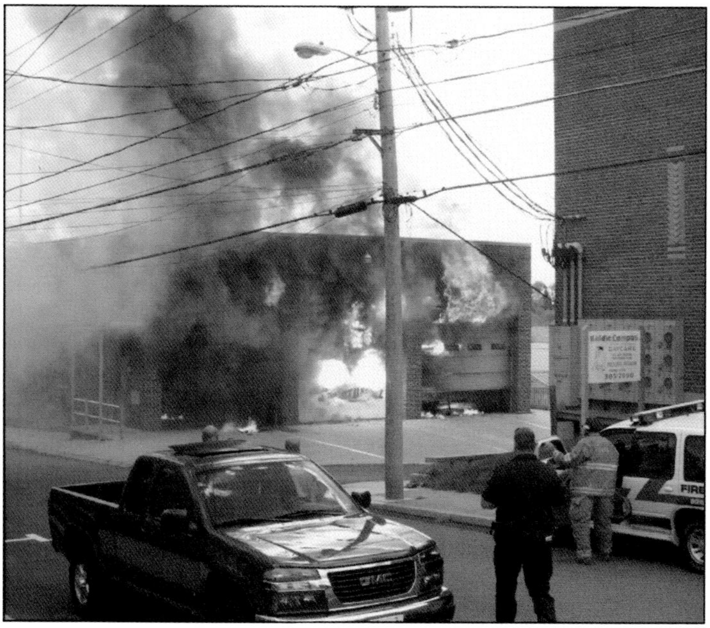

A heartbreaking scene greeted firefighters from the Schuylkill Hose Company No. 2, Schuylkill Haven, on September 12, 2007, as they arrived to find their firehouse heavily involved in fire. Destroyed in the fire were two American LaFrance pumpers, as well as the company's hazardous materials supply trailer. In the photograph, Schuylkill Haven assistant fire chief and longtime Schuylkill Hose member Glenn Sattizahn (right) establishes command. (Jim McGlaughlin.)

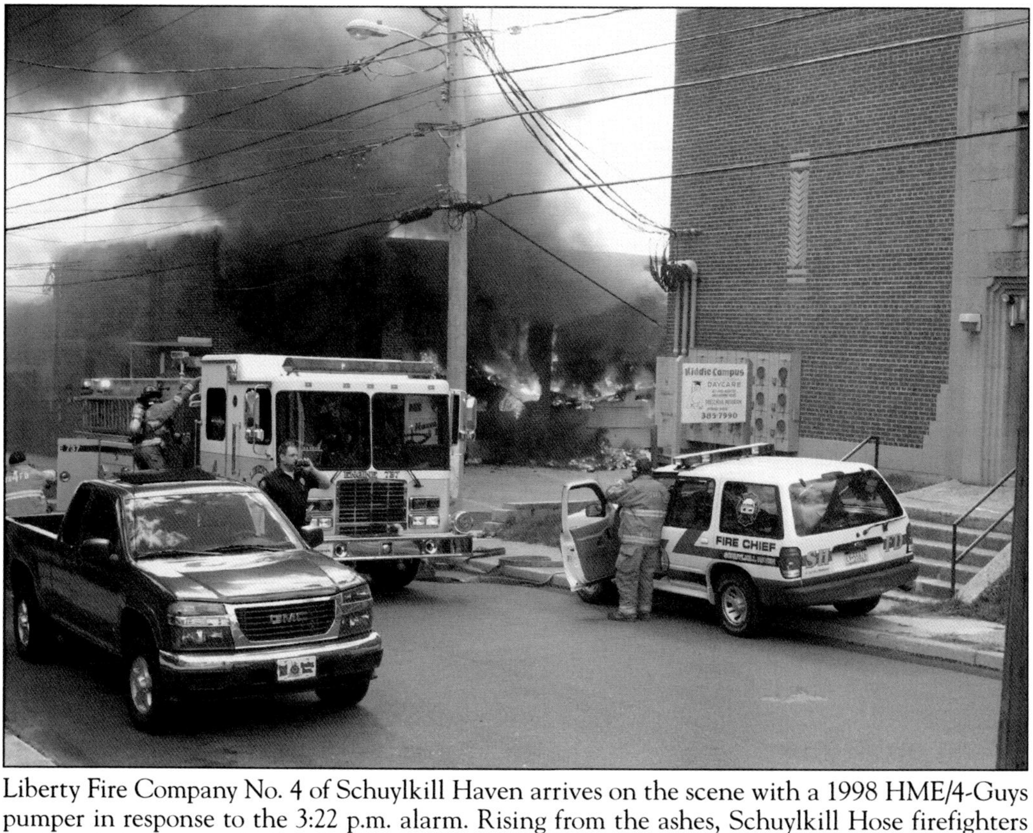

Liberty Fire Company No. 4 of Schuylkill Haven arrives on the scene with a 1998 HME/4-Guys pumper in response to the 3:22 p.m. alarm. Rising from the ashes, Schuylkill Hose firefighters dedicated their restored engine room, two new Pierce pumpers, and new utility/hazardous materials vehicle two years to the day and time of the initial alarm of fire. (Jim McGlaughlin.)

Three

East

Eastern Schuylkill County holds rich deposits of both anthracite and fertile farmlands. Both farming and mining communities developed in the early to mid-19th century. The borough of Tamaqua is the largest community in this area. The first fire company to organize in eastern Schuylkill County was the Reliance Hook and Ladder Company (later renamed American Hose Company) of Tamaqua in 1860. Tamaqua activated the Perseverance Hose and Steam Fire Company (renamed Citizens) in 1879 and added two more fire companies in the early 20th century. The South Ward Fire Company formed in 1913, and the East End Fire Company started in 1916.

Coaldale Borough, on the eastern edge of Schuylkill County, organized the Volunteer Fire Company No. 1 in 1906. The mining boroughs of Middleport and New Philadelphia also formed fire companies in the early 20th century. Middleport put together the Citizens Fire Company in 1903, and New Philadelphia started the Good Intent Fire Company in 1905. Schuylkill Township, located just west of Tamaqua, launched three fire companies in its three villages. The Brockton Fire Company formed in 1913, Tuscarora Fire Company in 1923, and the Mary-D Fire Company in 1950.

The farming areas of West Penn Township and Walker Township added fire protection in the 20th century. The Andreas Fire Company of West Penn Township began in 1942, and the West Penn Fire Company organized in 1998. The New England Fire Company of Walker Township assembled a fire company to provide fire protection to the farm-rich New England Valley. The New England Fire Company formed in 1953. Blythe Township organized two fire companies in the mining villages of Cumbola and Kaska. The Good Will Hose Company of Cumbola started in 1899, and the Kaska Fire Company came into being in 1957. The Ryan Township Fire Company organized in 1940 to increase fire protection to its rural area. Neighboring Rush Township added fire protection in 1941, with fire companies setting up in the villages of Quakake and Hometown.

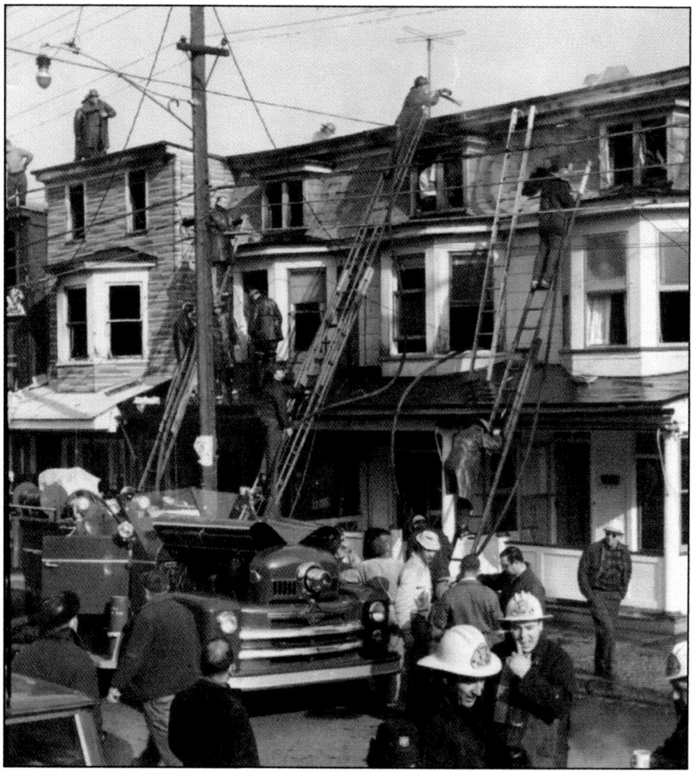

Tamaqua firefighters overhaul this fire in the 300 block of Washington Street on December 10, 1965. The alarm was sounded at 10:30 a.m. In all, six row homes were damaged and 21 individuals were left homeless by the fire. The Seagrave 70th Anniversary Series pumper of the American Hose Company of Tamaqua is in service in the photograph. (HSSC.)

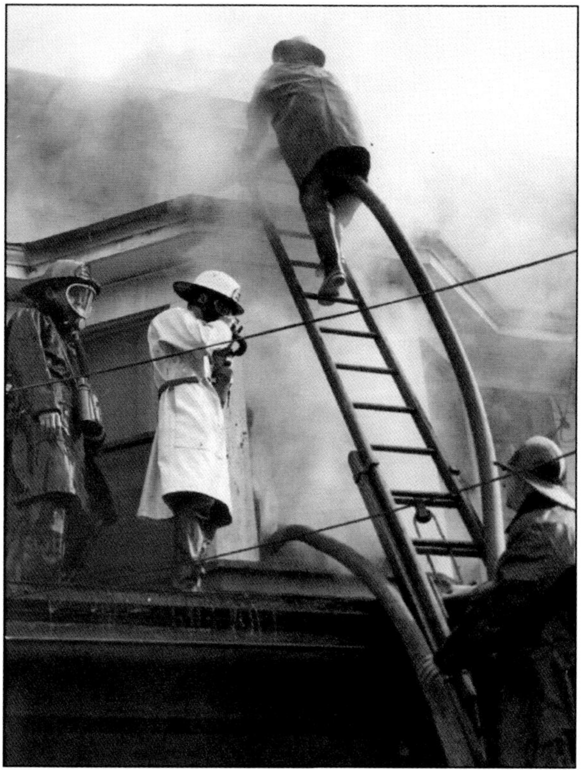

With some firefighters wearing filter-type masks, 2.5-inch hose lines are stretched to the upper floors of the frame homes in the 300 block of Washington Street in Tamaqua. The fire began in the second floor of 319 Washington. Tragically the occupants of the home in which the fire originated had lost their three young daughters to a fire on Brown Street eight years earlier. (HSSC.)

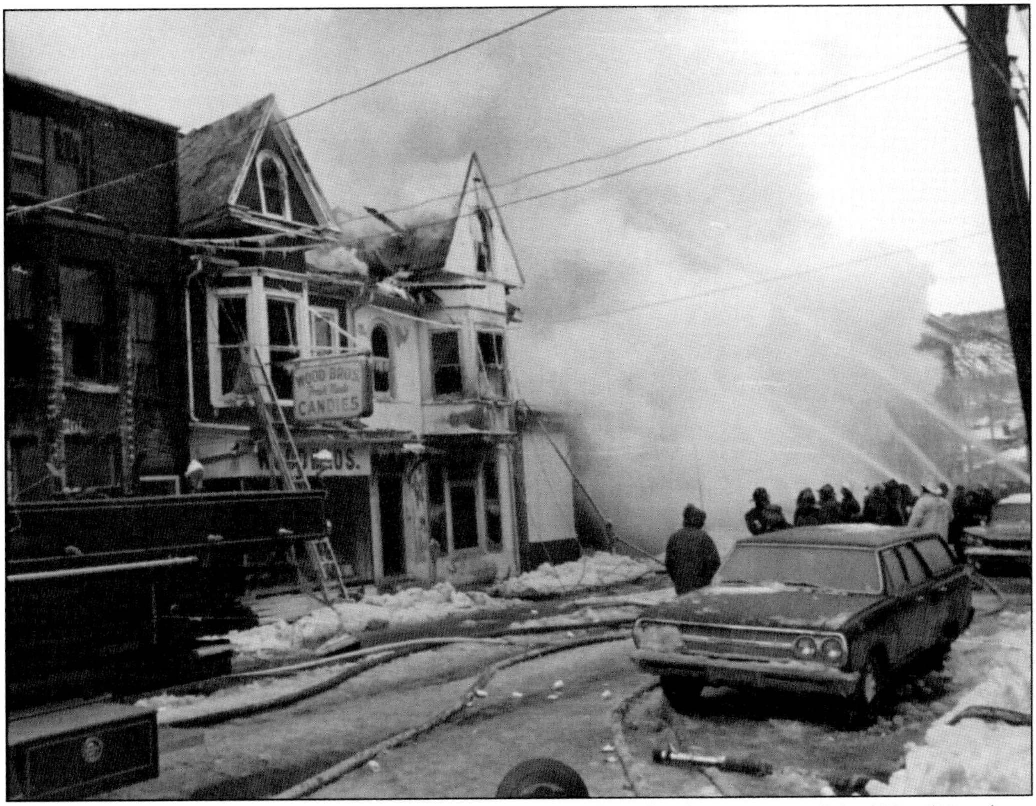

Fire originating in a chocolate-melting machine resulted in the destruction of the Wood Brothers Candy store, as well as almost the entire eastern side of the block on Mauch Chunk Street in Tamaqua. The alarm was sounded at 6:45 a.m. on February 21, 1972. Five firefighters were injured fighting the fire. Firefighters from Lansford, Coaldale, and New England (Walker Township) assisted Tamaqua. Thirty individuals were left homeless. (HSSC.)

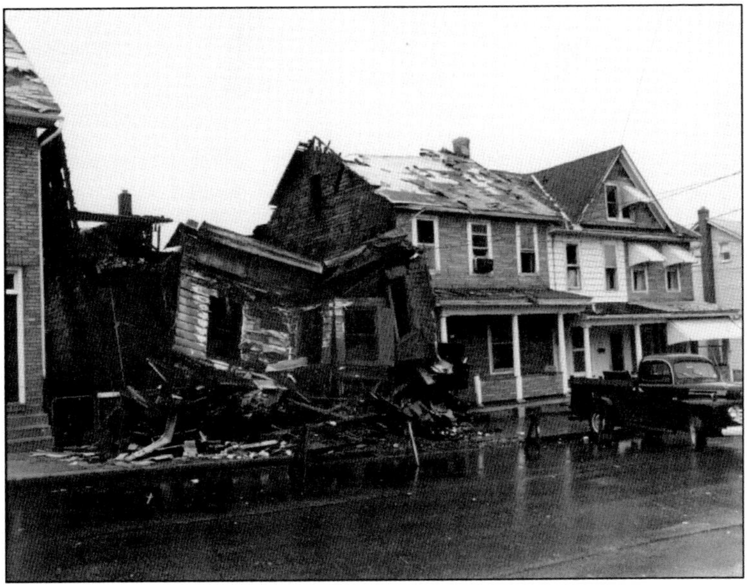

The remains of a tavern in Tamaqua's Dutch Hill section are pictured here following a devastating fire on March 11, 1973. The alarm was sounded at 4:35 a.m. Eleven men were injured, and 16 individuals were left homeless. The fire originating in the tavern on East Union Street spread to three row homes to the west (right) and a grocery store to the east (left). (HSSC.)

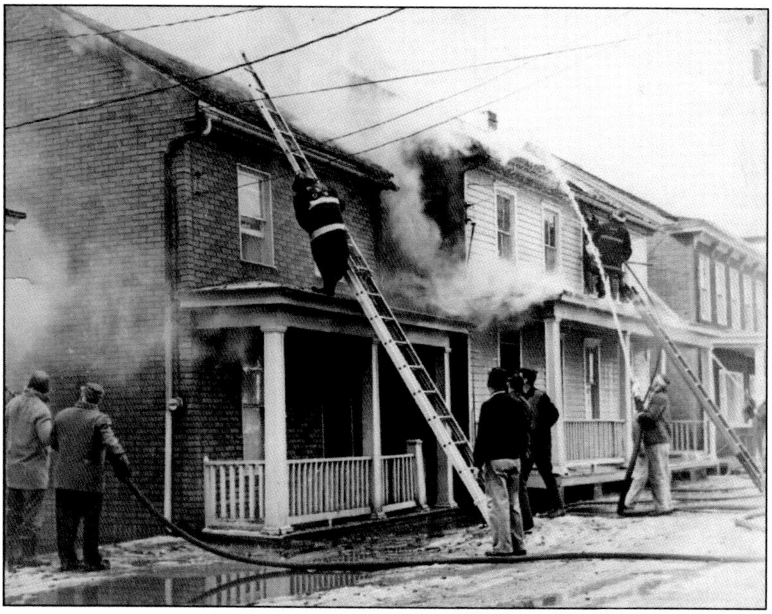

Ten individuals were left homeless after this fire on Water Street in Cumbola destroyed three frame dwellings on February 22, 1966. Firefighters from New Philadelphia, Port Carbon, and Pottsville assisted Cumbola in battling the 10 a.m. blaze. (SHFS.)

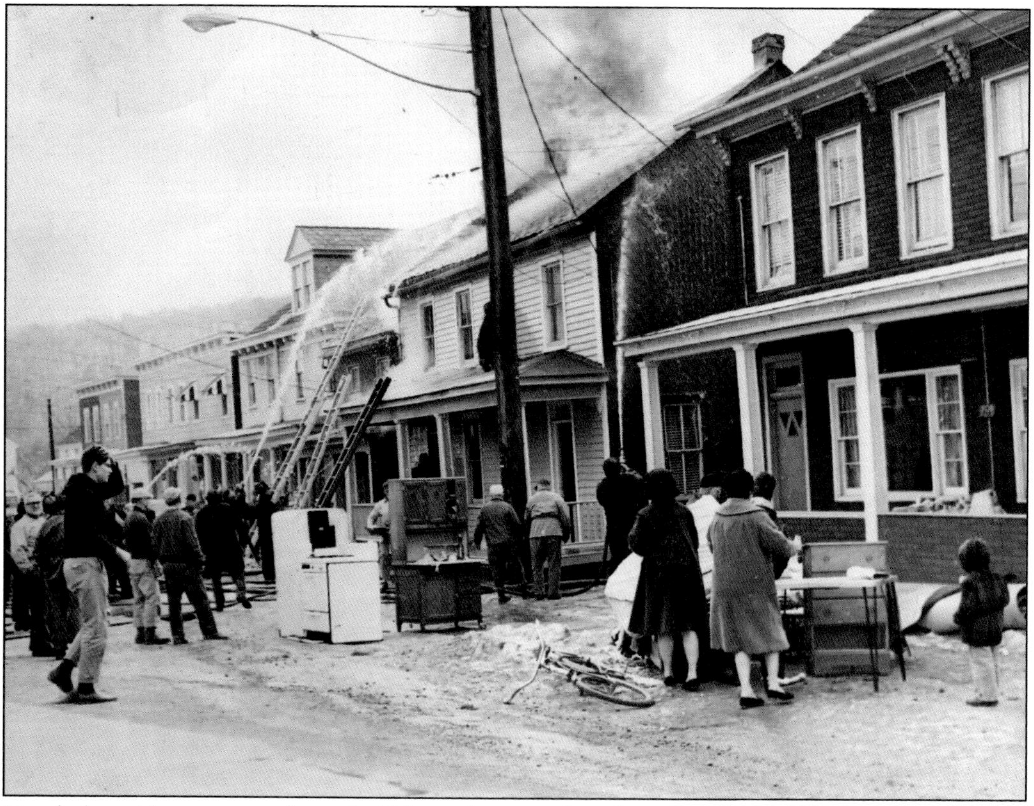

Cumbola residents remove household furnishings and appliances from neighboring homes as fire consumes the roof of a frame double home in the center of the photograph. Firefighters obtained water to fight the fire from a nearby fire hydrant, as well as a creek running to the rear of the homes. Temperatures were in the 20s when the fire occurred on February 22, 1966. (HSSC.)

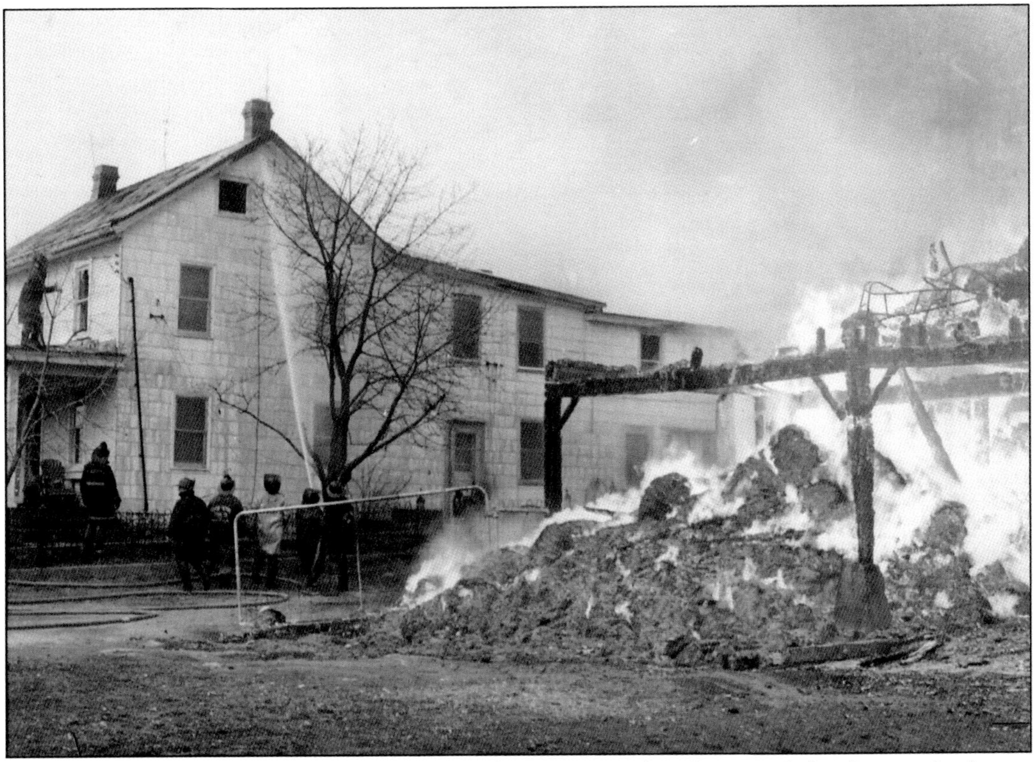

On Sunday, March 8, 1964, at approximately 1:00 p.m., fire destroyed this barn, which was leased by Guers Dairy on Tumbling Run Road in Blythe Township. Firefighters work to save the frame home, which was seriously exposed to the fire. The home suffered minor fire damage. Firefighters from Cumbola, New Philadelphia, New Ringgold, Middleport, and Pottsville fought the blaze. (SHFS.)

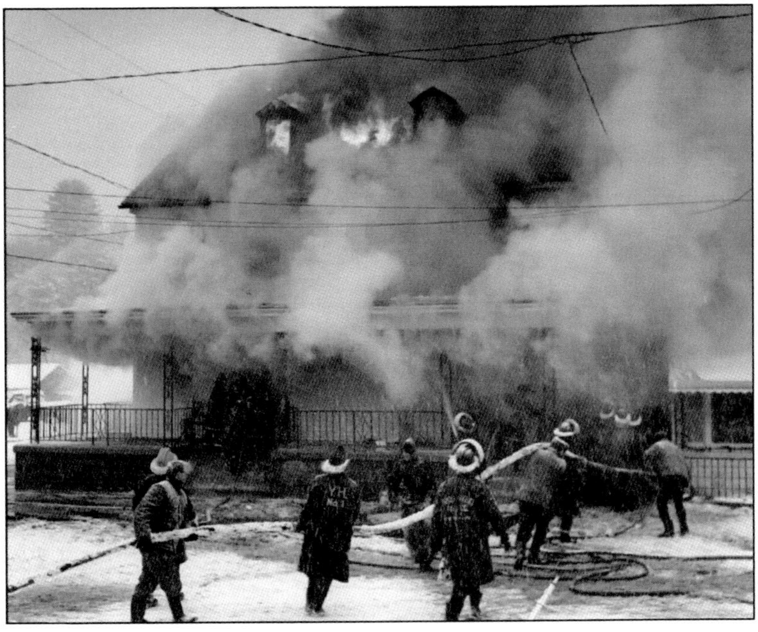

At 6:15 a.m. on New Year's Day in 1971, fire destroyed Mazloom's Restaurant and Bar along with a beauty salon and apartments at 59 Valley Street in New Philadelphia. The double-block frame building was over 100 years old at the time of the blaze. Firefighters from Kaska and three Pottsville fire companies assisted at the fire, which occurred at 6:15 a.m. during a snowstorm. (SHFS.)

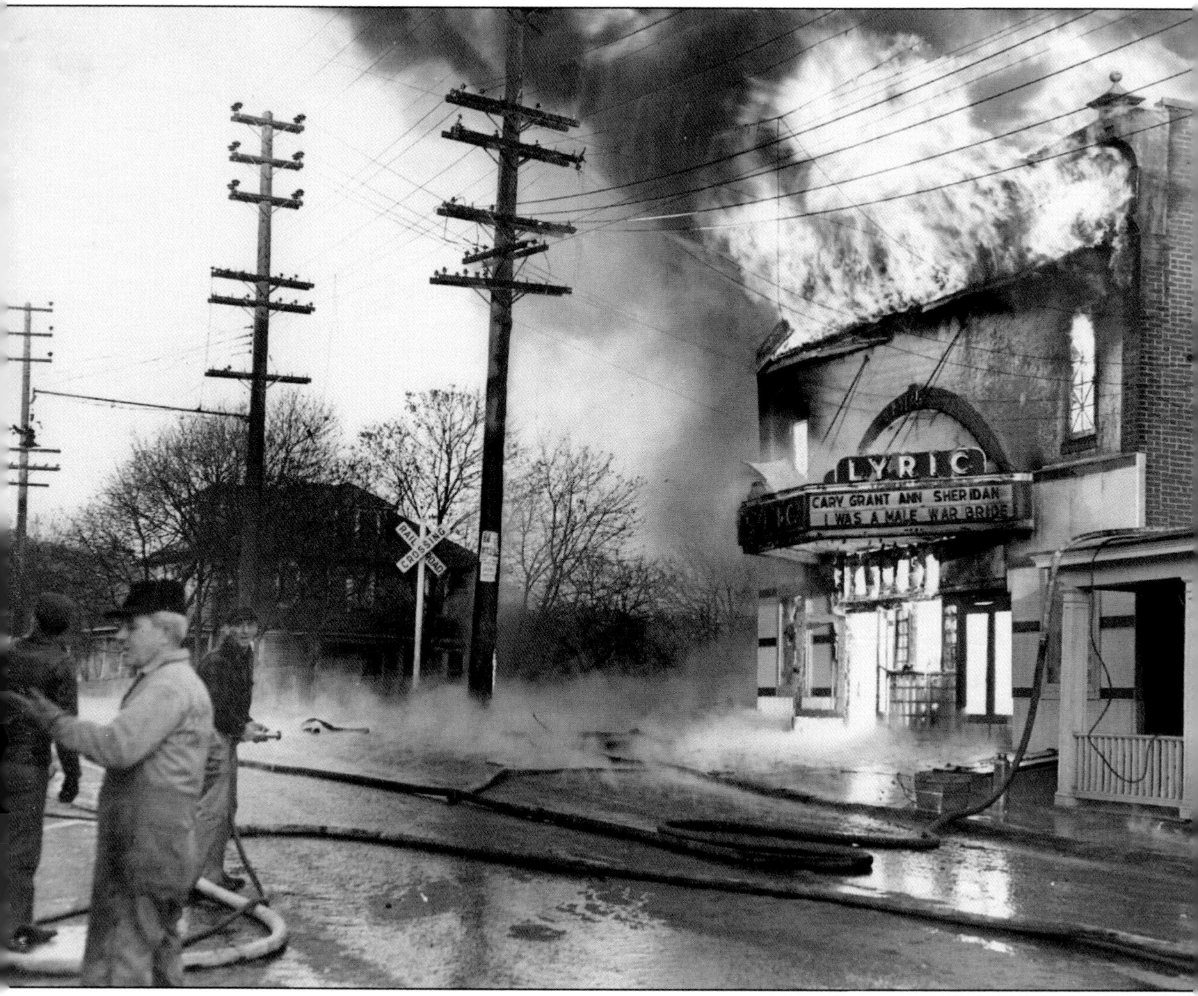

I Was a Male War Bride with Cary Grant was showing when fire struck the Lyric Theater in New Philadelphia at approximately 3:00 p.m. on December 16, 1949. The theater was built in 1924. Firefighters struggled to keep the fire from extending to the home at the right of the theater as well as eight frame homes to the rear. Five fire companies were on the scene, including New Philadelphia, Cumbola, Port Carbon, Citizens of Palo Alto, and the American Hose Company of Pottsville. Damage to the theater was estimated at $75,000. (HSSC.)

Christmas morning in 1982 dawned tragically in the borough of Tamaqua. At 4:00 a.m., firefighters were called to this home at 117 West Union Street. They arrived to find the three-story frame dwelling heavily involved in fire. A mother and her three young children perished in the blaze, which originated in the first-floor middle room. The homes at 115 West Union and 121 West Union were also damaged. (HSSC.)

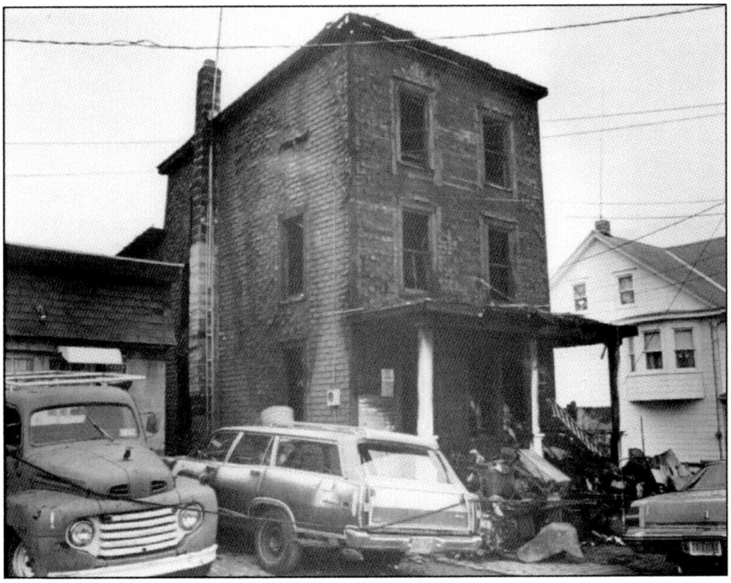

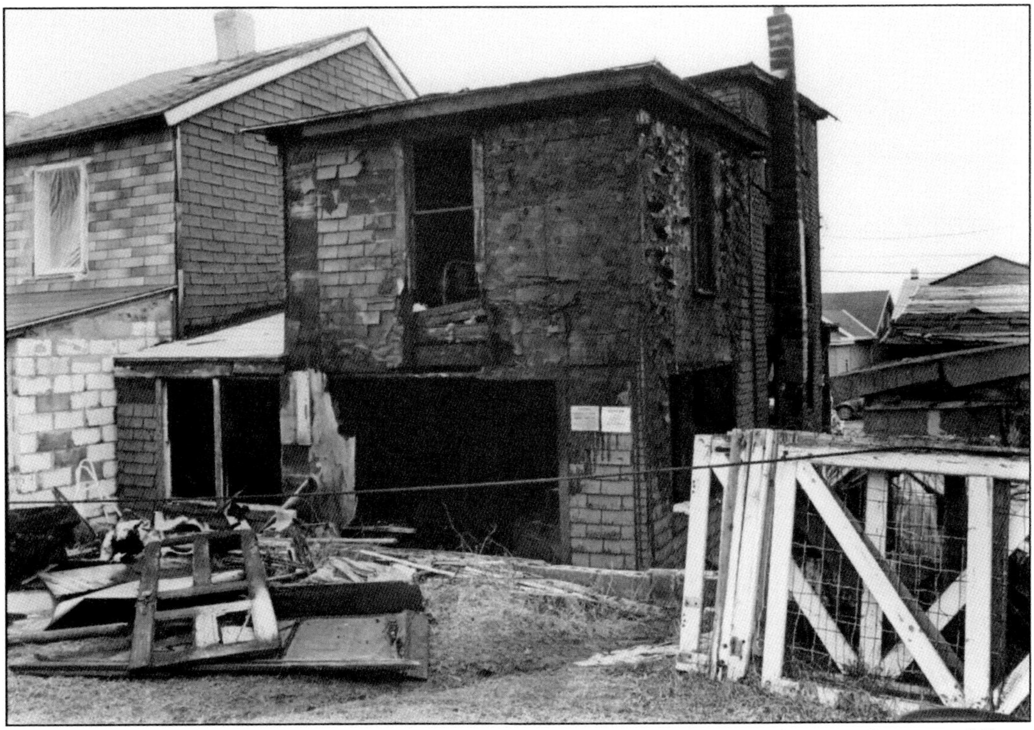

This photograph shows the rear of the Tamaqua home. The family was asleep on the second floor when the parents were awakened by crackling. When the parents opened the bedroom door, they were hit by smoke and flames spreading rapidly up the interior stairwell. Cut off from the children, the father escaped from a second-floor window. The mother was unable to get out. (HSSC.)

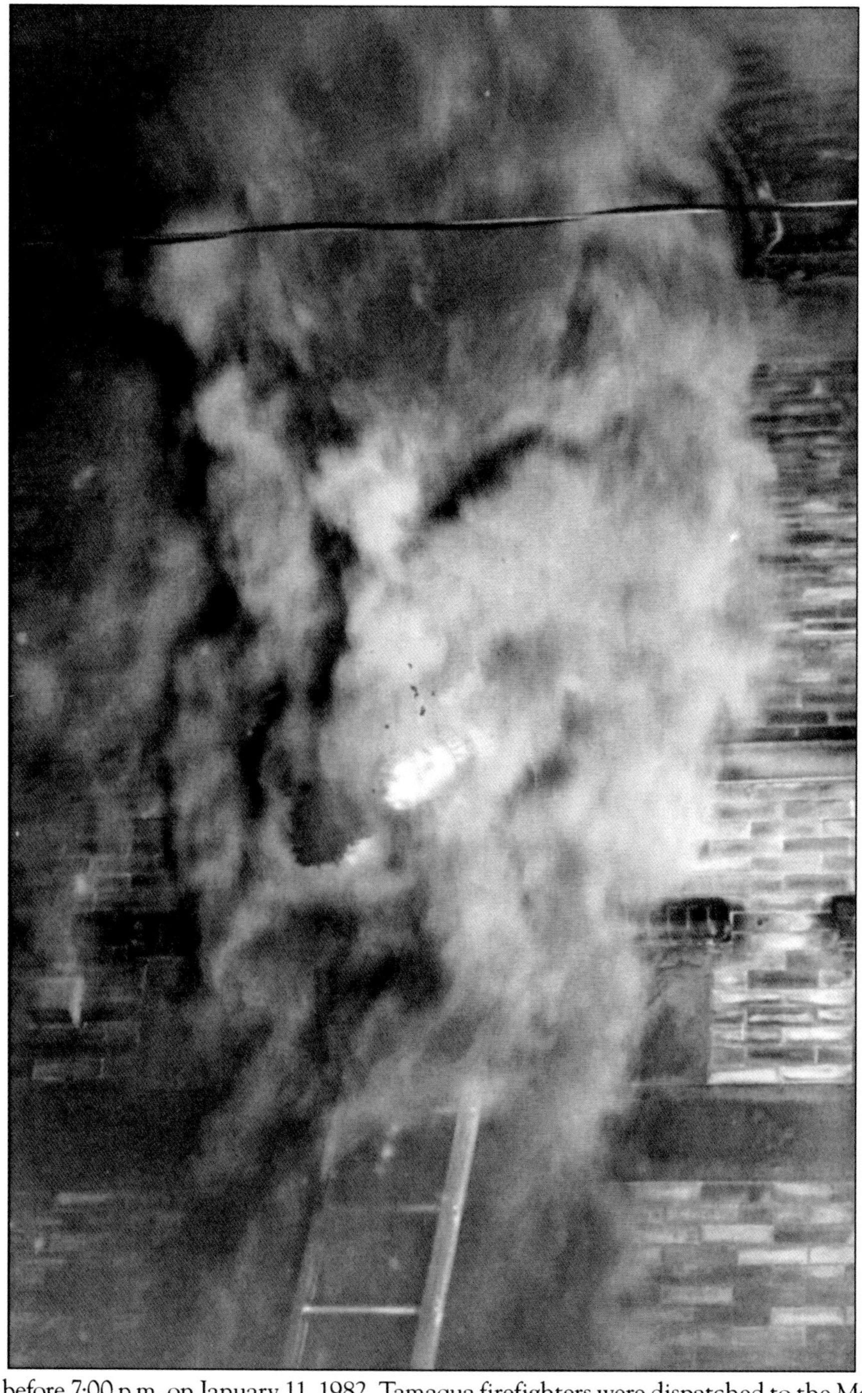

Shortly before 7:00 p.m. on January 11, 1982, Tamaqua firefighters were dispatched to the McAloose Building at 14 West Broad Street. Arriving on the scene, they learned of an occupant in the third-floor apartment unaccounted for and began a primary search of the floor in heavy smoke conditions. As their air supplies depleted, they were forced to the third-floor windows. In the photograph, firefighter John O'Brien hangs out the third-floor window above the ladder with just his head and SCBA cylinder visible through the thick smoke. (HSSC.)

The situation at the McAloose Building in Tamaqua on January 11, 1982, becomes desperate. Trying to right himself on the ladder, firefighter John O'Brien fell. His fall was slowed by the grasp of another man on the ladder, who was coming to his aid. Firefighters on the ground also cushioned his descent. O'Brien was admitted to the hospital suffering from smoke inhalation. (HSSC.)

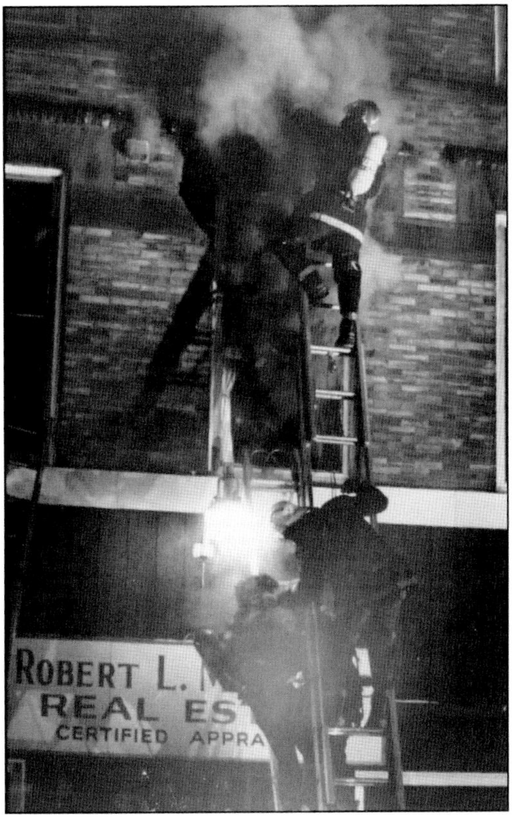

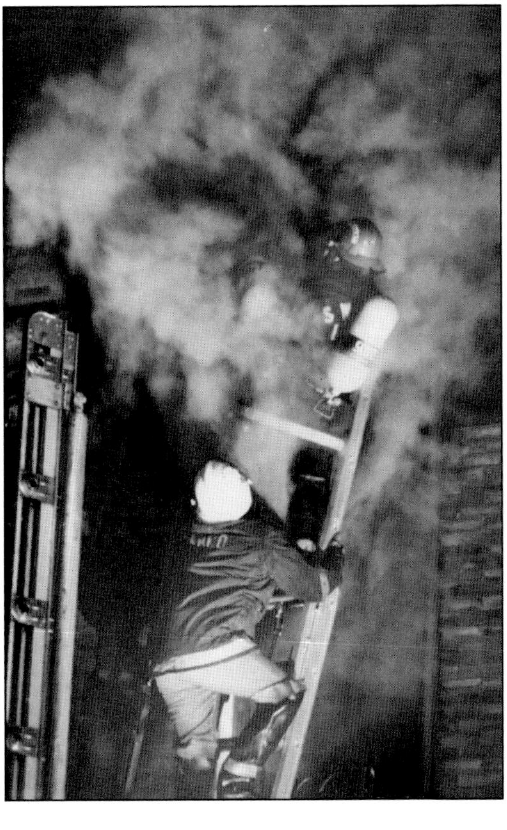

Another ladder is placed to the McAloose Building as comrades assist firefighter John Fredericks from the third-floor window after his air supply became depleted. The top of Fredericks's SCBA cylinder is visible to the left of the man on the ladder. The occupant for whom they were searching was found uninjured in the rear of the third floor by firefighters entering from the rear of the building. (HSSC.)

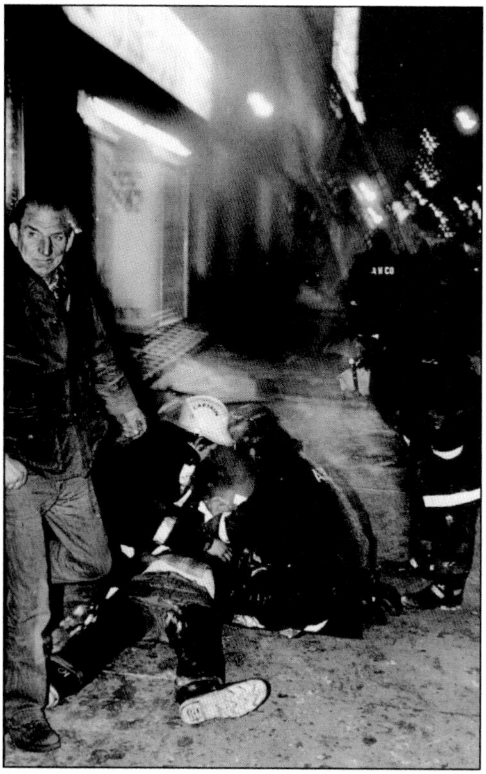

Colleagues attend to firefighter Fredericks, lying on the sidewalk, after he fell trying to exit the third-floor window onto the ladder. It was believed that he was struck by an object, causing the fall. He was admitted to the hospital suffering from smoke inhalation. In all, four firefighters were injured fighting the fire. (HSSC.)

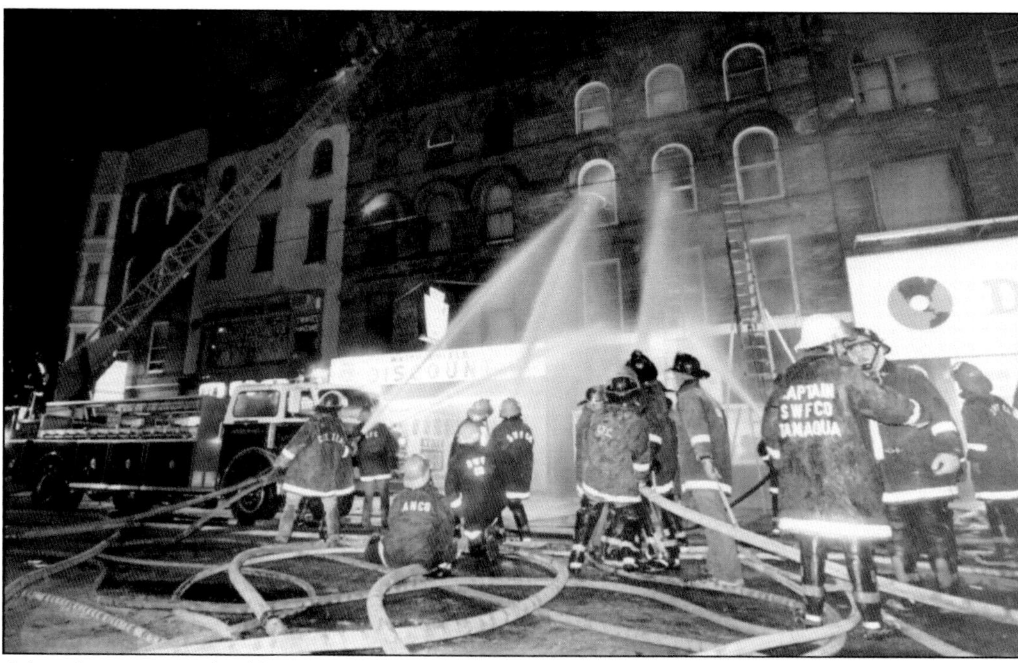

After the occupant had been removed and the dramatic rescue of the firefighters from the third floor was complete, hose streams were directed into the McAloose Building from the exterior. Tamaqua's 1978 American LaFrance 100-foot aerial ladder has its "main" placed to the roof. It was believed that the fire originated in a first-floor laundromat. (HSSC.)

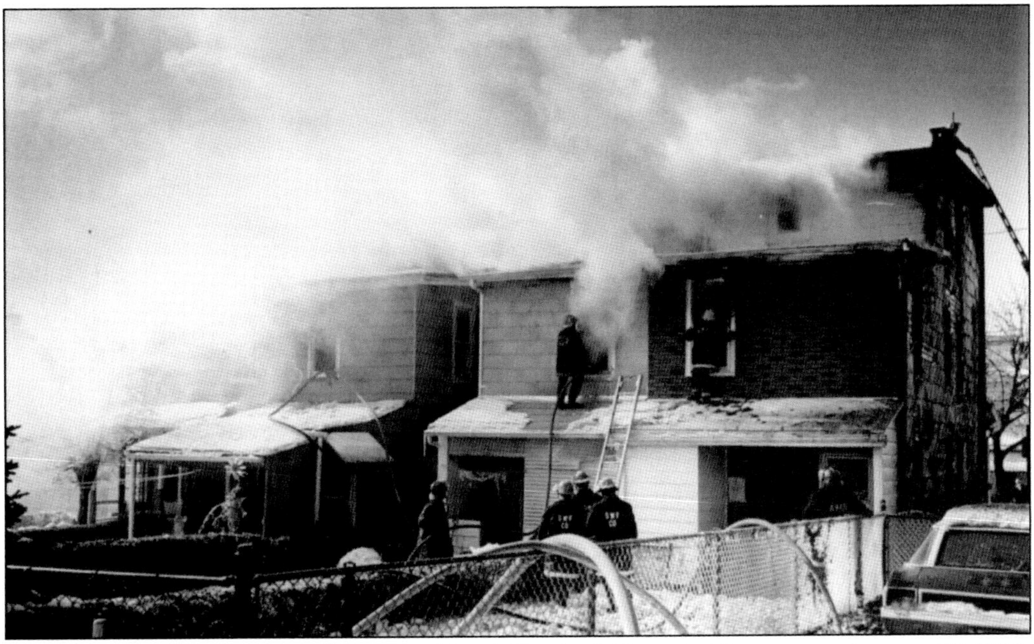

Firefighters work in the rear of homes on North Elizabeth Street in Tamaqua on January 26, 1982. The homes at the end of the row, 315 and 313 North Elizabeth, were gutted, while the other two homes in the row, 309 and 311, were heavily damaged. In all, nine individuals were left homeless. The damage totaled $100,000. (HSSC.)

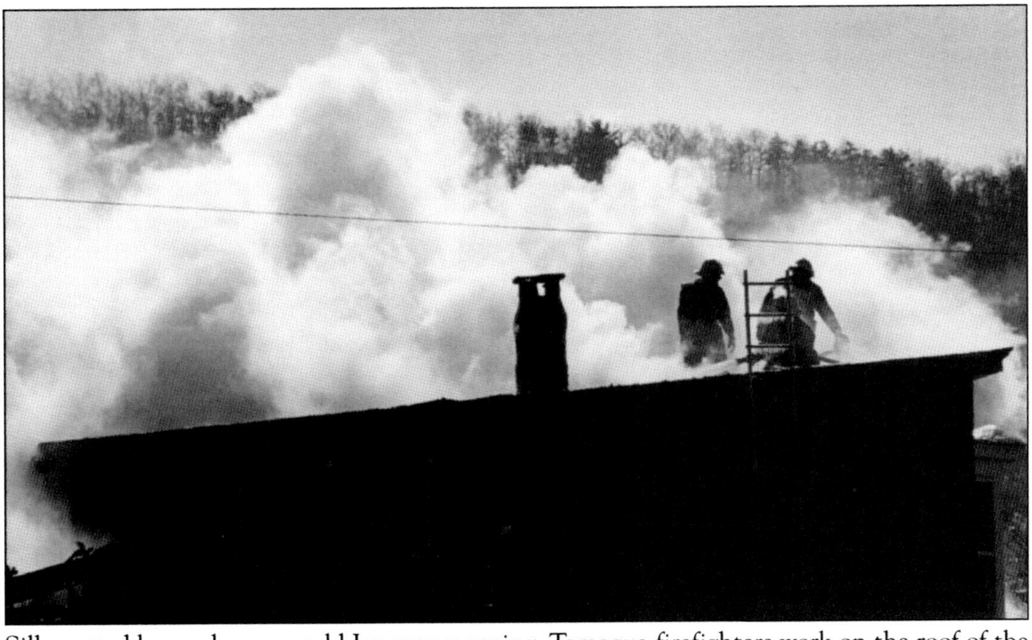

Silhouetted by smoke on a cold January morning, Tamaqua firefighters work on the roof of the North Elizabeth Street homes on January 26, 1982. Firefighters remained on the scene throughout the day. All four borough fire companies fought the inferno. This was the third serious fire to strike the borough in a three-month period. (HSSC.)

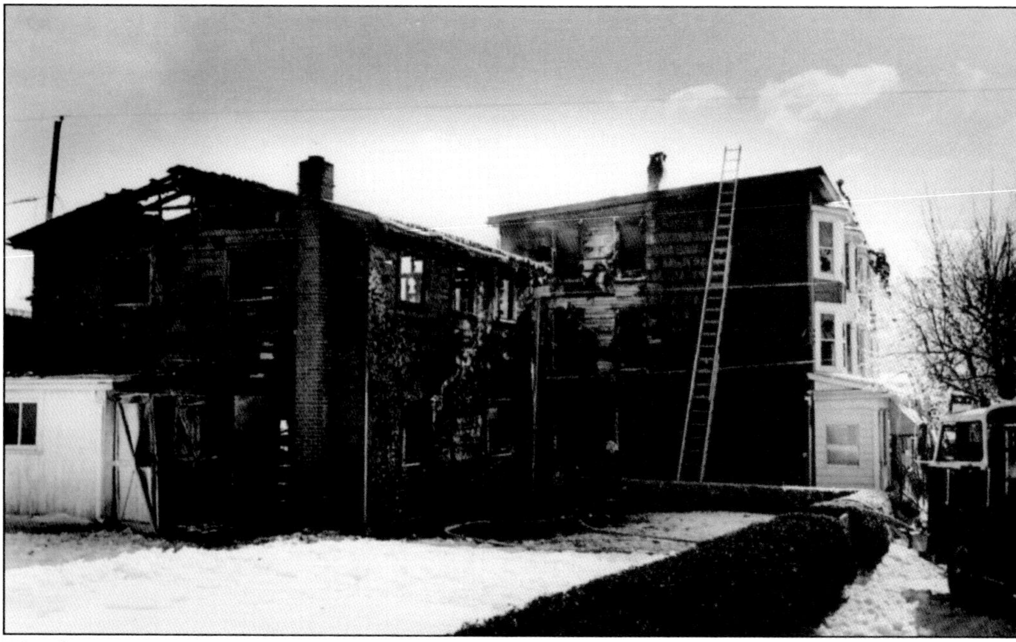

The fire originating in the garage in the Dutch Hill section of Tamaqua on January 26, 1982, spread rapidly to the row of frame homes at 309–315 North Elizabeth Street. To the extreme right of the photograph is Tamaqua's 1978 American LaFrance 100-foot aerial ladder truck nosed into the scene. (HSSC.)

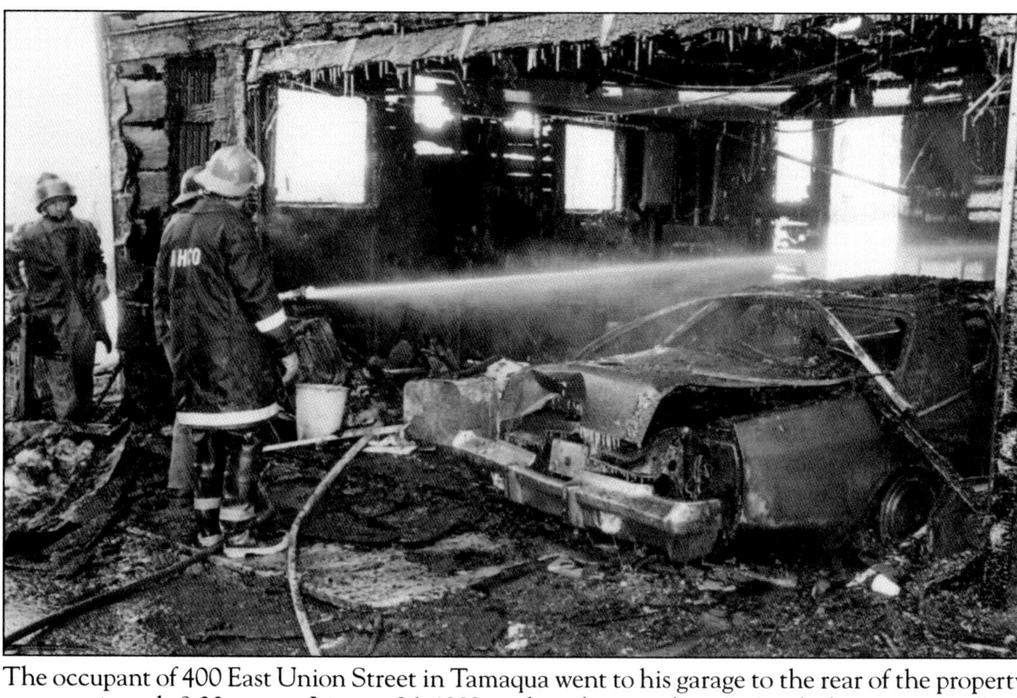

The occupant of 400 East Union Street in Tamaqua went to his garage to the rear of the property at approximately 9:00 a.m. on January 26, 1982, and tried to start his car. Smoke began to pour from the front of the vehicle, and the individual ran to transmit the alarm. By the time he returned, both the car and the garage were well involved. (HSSC.)

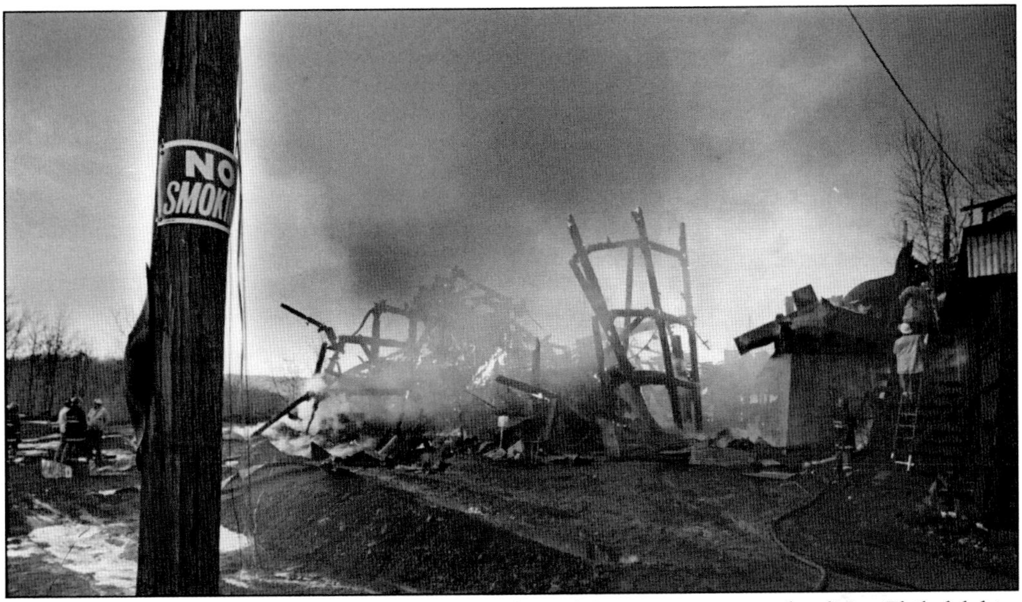

On December 23, 1978, fire destroyed the Shamrock Coal Company just outside of New Philadelphia. It had formerly been the Silver Creek Coal Company. As can be seen in the photograph, the breaker was completely destroyed. Responding to the 2:05 p.m. alarm were firefighters from New Philadelphia, Orwigsburg, Mary-D, Cumbola, Middleport, and Landingville. The "No Smoking" sign seems ironic! (HSSC.)

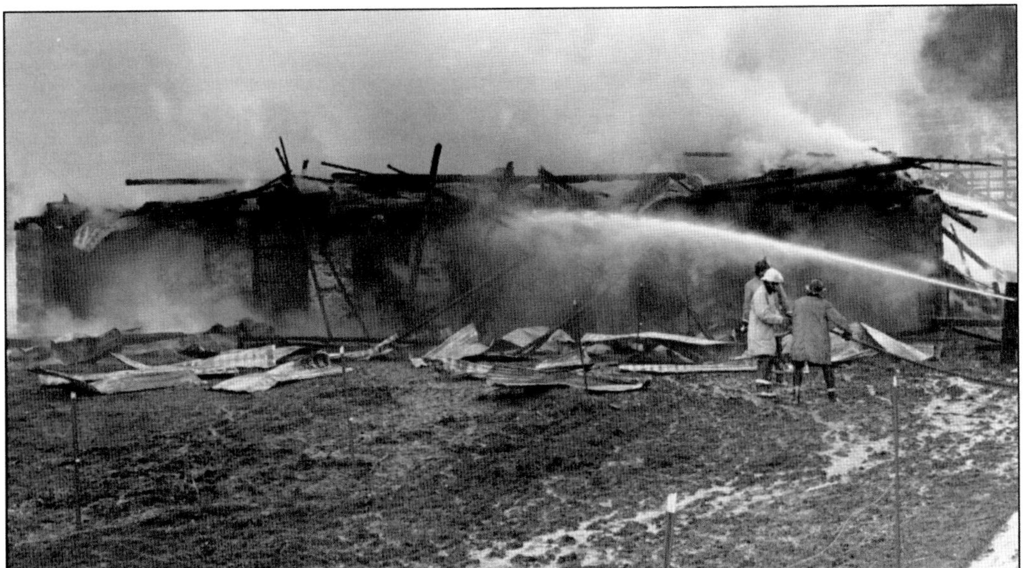
This 19th-century barn on the Leiby Farm in Walker Township burned to the ground on April 16, 1983. Firefighters arrived in response to the 2:40 p.m. alarm to find the structure fully involved in fire. Seven fire companies fought the blaze in gusty wind conditions. (HSSC.)

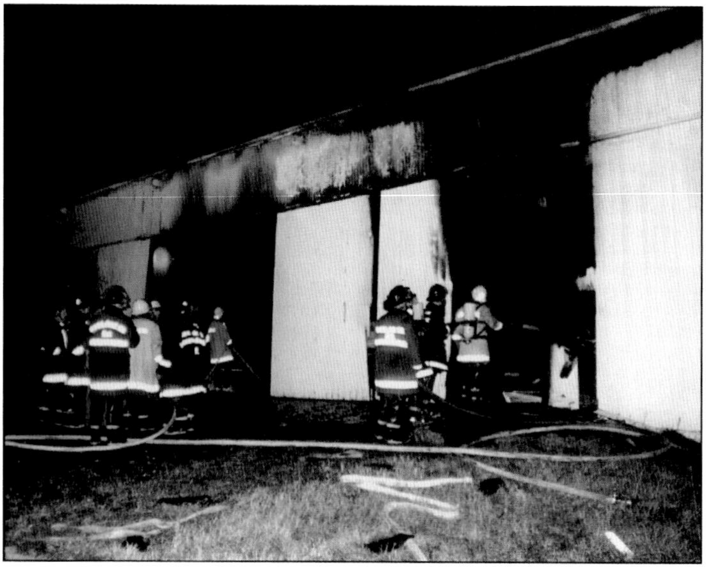

Fire struck the Reynolds Wood Products facility in Walker Township off Route 443 on October 23, 1984. Firefighters arriving at 6:30 p.m. were confronted with heavy fire conditions in the building. Seven employees were left temporarily without work. The fire was believed to have been electrical in nature. Firefighters from New England (Walker Township), Andreas, New Ringgold, and Mahoning Valley (Carbon County) fought the flames. (HSSC.)

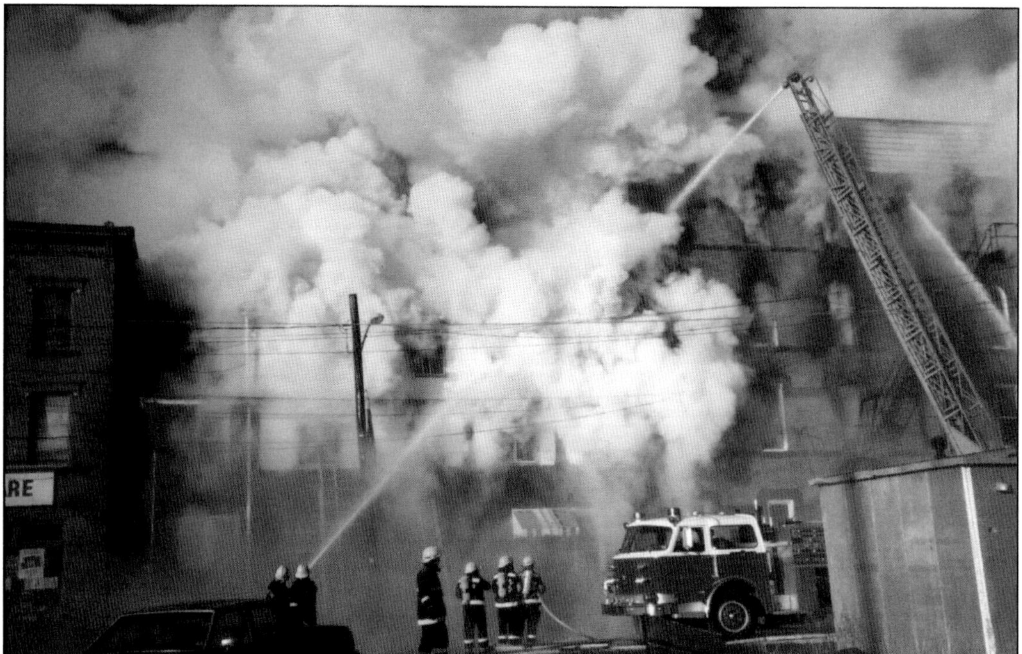

Tamaqua's American LaFrance 100-foot aerial ladder truck directs a stream into the upper floors of the Tamaqua Hotel on Sunday, December 27, 1992. Firefighters arrived to find heavy fire on the second and third floors of the four-story structure at 8 West Broad Street. One female tenant was killed as she went through the building trying to warn others. (Authors' collection.)

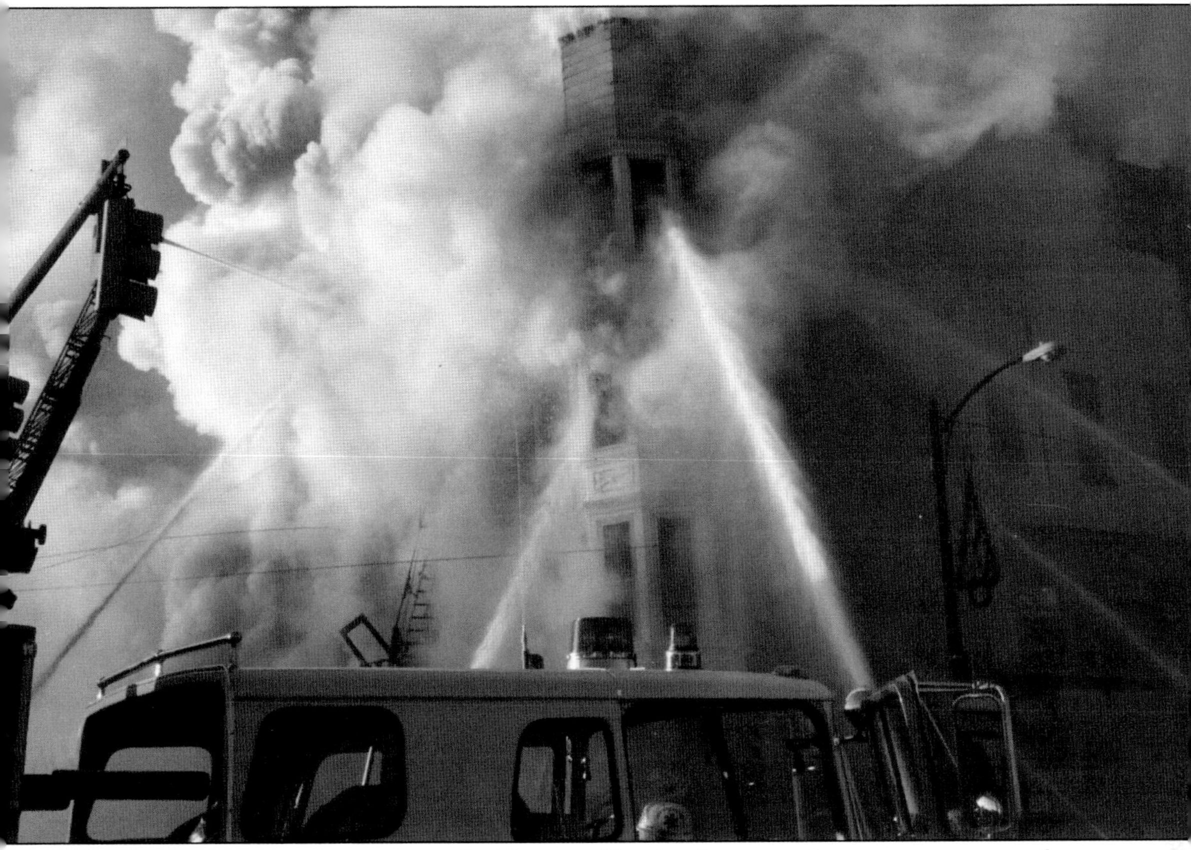

Numerous streams are directed into the Tamaqua Hotel as heavy smoke continues to pour from the building. While the hardware store to the right was damaged, firefighters were successful in preventing the fire from extending further down the block in the heart of Tamaqua's business district. The fire began at approximately 8:00 a.m. on a very cold December morning. The American Fire Company in nearby Lansford, Carbon County, assisted with an aerial ladder truck. The cab roof of the American Hose Company of Tamaqua's 1981 Ford/Darley pumper is visible in the photograph. (Authors' collection.)

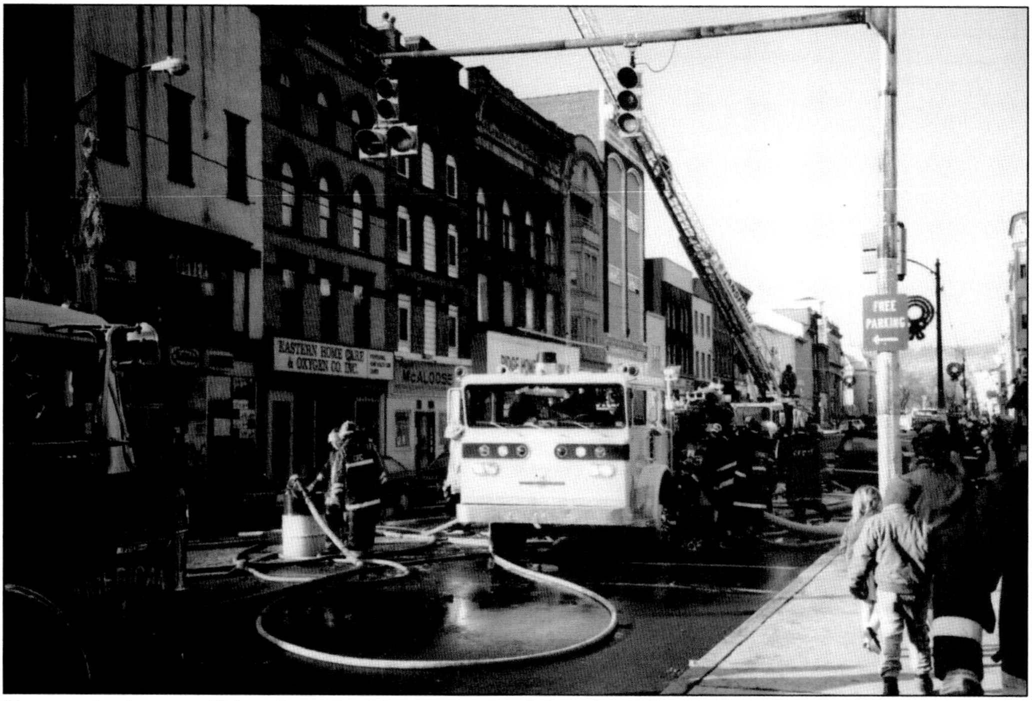

During the long, cold battle at the Tamaqua Hotel fire, firefighters use an empty drum to support their 2.5-inch hose line. Pictured at left is the American Hose Company of Tamaqua's 1981 Ford/Darley pumper, and to the right is the Citizens Fire Company of Tamaqua's 1978 American LaFrance pumper. (Authors' collection.)

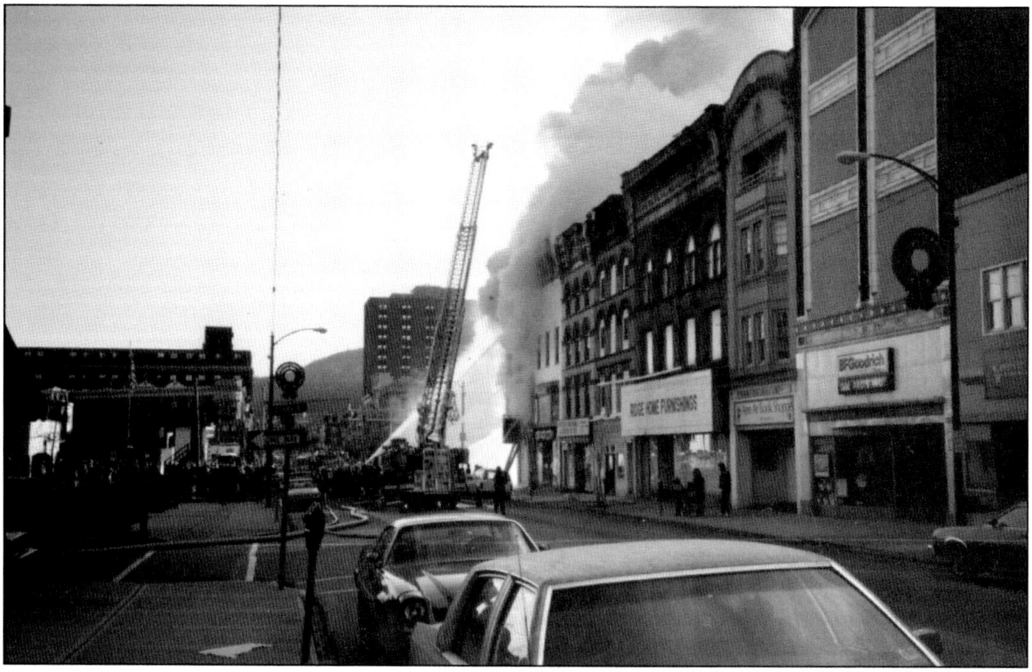

This is an overall view of the scene looking east on Broad Street in Tamaqua on December 27, 1992. The American Fire Company of Lansford is in service with their E-One 75-foot quint.

Four

West

Minersville's Mountaineer Hose Company, or "Mounties," was the first fire company organized in western Schuylkill County. This brigade formed just prior to the Civil War in 1861. Minersville soon added three additional fire companies to bolster protection for its coal-mining community. The Good Will formed in 1867, the Independent organized in 1869, and the Rescue Hook and Ladder started in 1871. Minersville's youngest fire company, the New Minersville Fire Company, formed in 1926.

The Tremont Fire Company No. 1 commenced in Tremont Borough in 1878. Further west, the community of Tower City also activated a fire company in the late 19th century. The Volunteer Fire Company No. 1 of Tower City formed in 1899. The adjoining township of Porter assembled five fire companies in the 20th century to protect the township's villages. The Orwin Fire Company organized in 1920, followed by the Reiner City Hook and Ladder Company in 1922. This company was later renamed the Muir Fire Company. The villages of Reinerton and Sheridan both initiated fire companies in 1926. The last Porter Township fire company to organize was the Joliett Fire Company in 1956.

The farm-rich Hegins Valley acquired fire protection in the 1920s. The Hegins Fire Company started in 1923, and the Valley View Fire Company formed in 1927. The neighboring townships of Mahantongo and Eldred augmented protection in their farming areas by introducing companies in the 1950s. The Mahantongo Fire Company of Pitman commenced in 1951, and the Sacramento Fire Company of Eldred formed in 1959. Just over the mountain in Frailey Township, the Donaldson Fire Company organized in 1926.

The coal-rich townships of Branch, Cass, Foster, and Reilly inaugurated several fire companies in the early to mid-20th century. Citizens of Branch Township created the Good Intent Fire Company of Llewellyn in 1908 and later the Phoenix Park Fire Company in 1966. Cass Township added the Clover Fire Company of Heckscherville in 1938 and the Duncott Fire Company in 1951. In neighboring Foster Township, the Mackeysburg Fire Company organized in the 1930s but disbanded in the mid-1960s. The Mount Pleasant Fire Company of Foster Township formed in 1947. Reilly Township added two fire companies to protect its villages. The Citizens Fire Company of Branch Dale originated in 1920, and the Newtown Fire Company of Zerbe started in 1954.

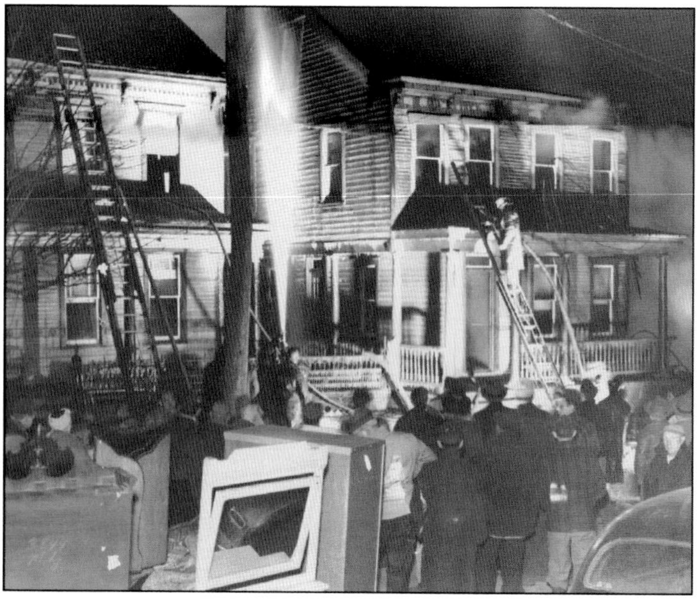

In anticipation of fire spreading further, furnishings of homes exposed to fire have been removed and are pictured in the foreground. Eleven individuals were left homeless in what became known as the Starr Store Fire in Llewellyn on February 2, 1953. Fire companies from Llewellyn, Minersville, South Cass, Heckscherville, Forestville, Branchdale, and Pottsville fought the blaze. (Good Intent Hose Company, Llewellyn.)

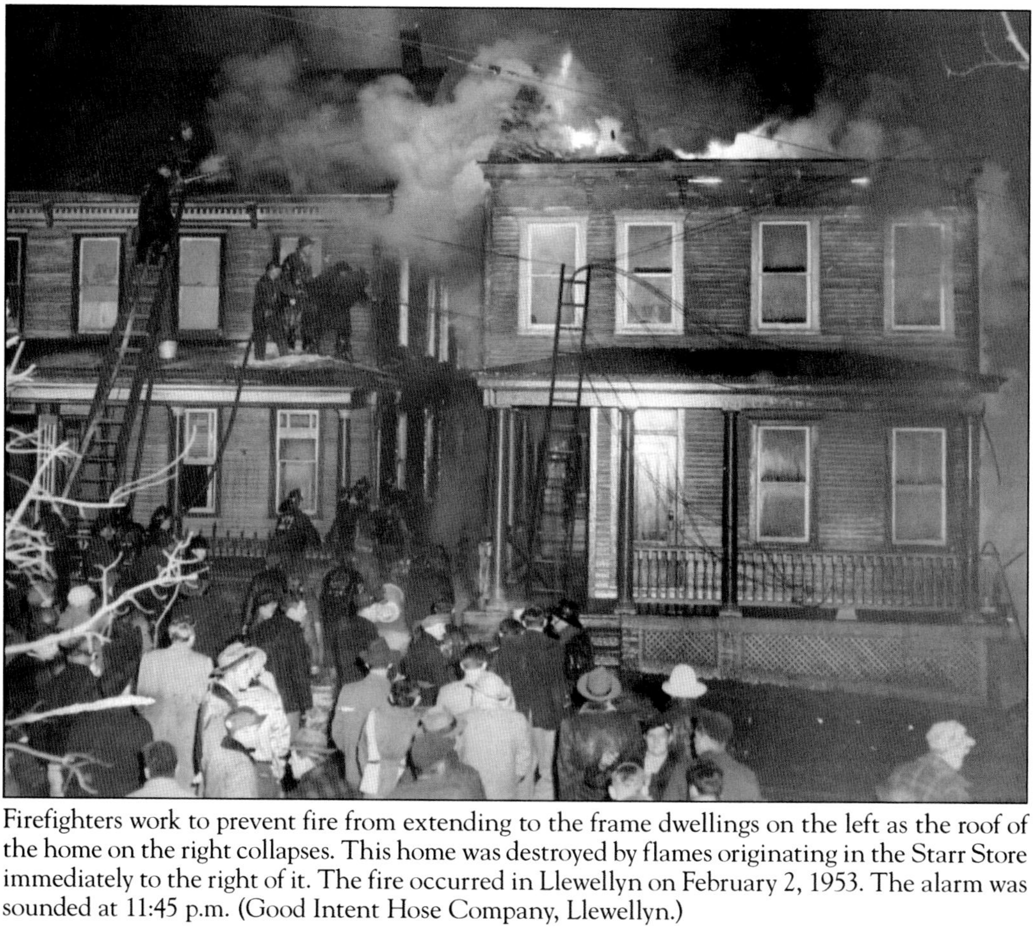

Firefighters work to prevent fire from extending to the frame dwellings on the left as the roof of the home on the right collapses. This home was destroyed by flames originating in the Starr Store immediately to the right of it. The fire occurred in Llewellyn on February 2, 1953. The alarm was sounded at 11:45 p.m. (Good Intent Hose Company, Llewellyn.)

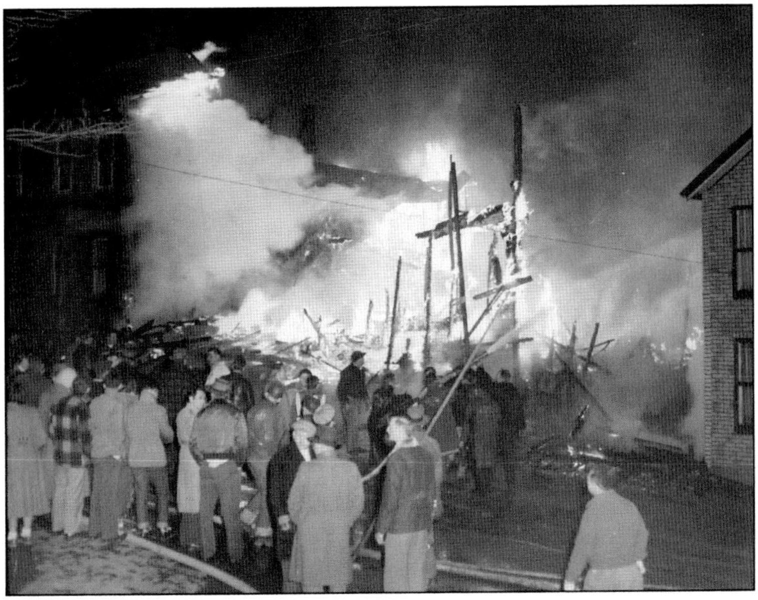

The Starr Store on Willing Street in Llewellyn was heavily involved in fire when firefighters arrived just before midnight on February 2, 1953. The building eventually collapsed, with much of it coming down into the street. Firefighters work to prevent the blaze from extending to the frame building on the right. Two firefighters were injured. (Good Intent Hose Company, Llewellyn.)

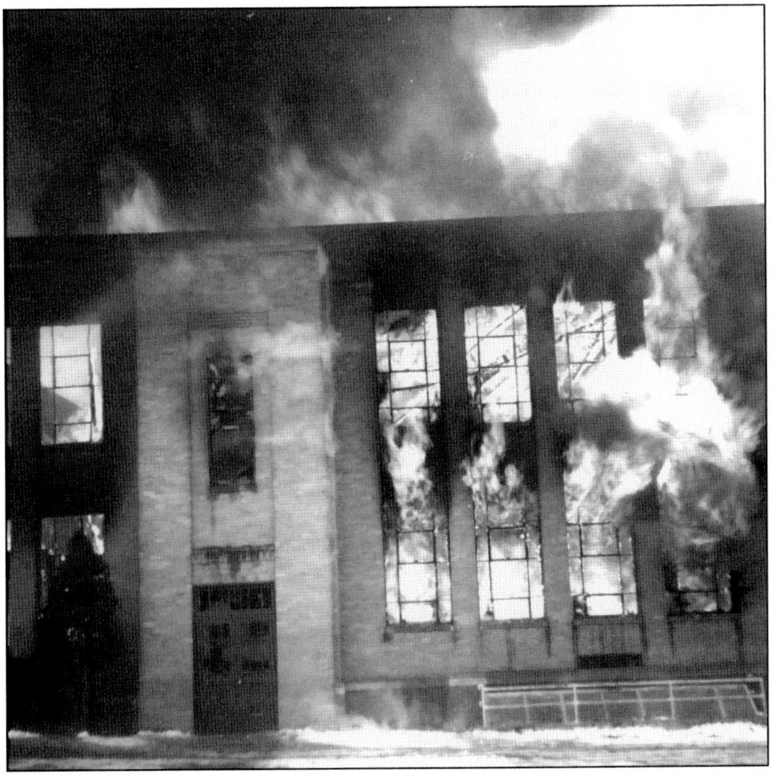

Flames engulf the Llewellyn Elementary School on December 14, 1963. The building, formerly the Branch Township High School, was and is part of the Minersville Jointure—currently the Minersville Area School District. The structure was rebuilt as a one-story school and is currently home to the Llewellyn Early Childhood Education Center. Fire companies from eight communities fought the fire. (Herb Spangler via Good Intent Hose Company, Llewellyn.)

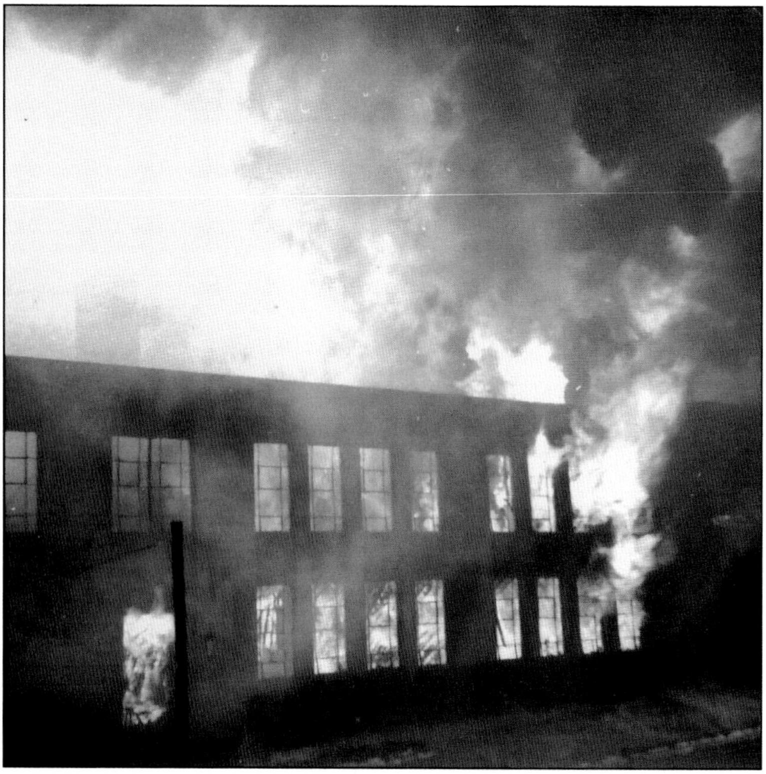

Along the Taylor Street side of the former Llewellyn Elementary School, the interior collapses as fire consumes the building. The alarm was sounded at 9:30 a.m., and within an hour, the classroom portion of the building was doomed. Thankfully the fire occurred on a Saturday when school was not in session. (Herb Spangler via Good Intent Hose Company, Llewellyn.)

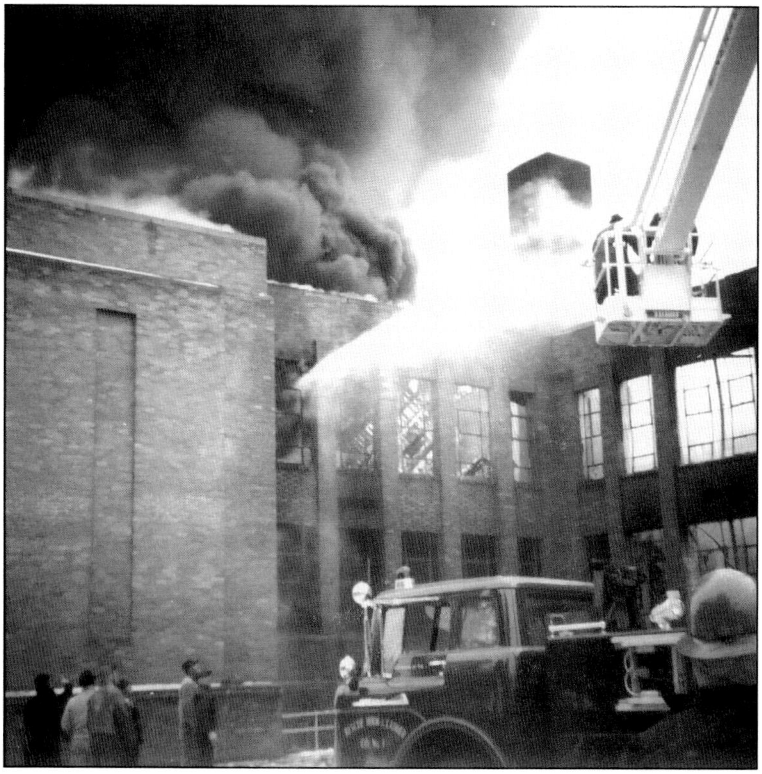

The Rescue Hook and Ladder Company of Minersville had just received a 65-foot Ford/Pitman snorkel in October 1963. In the photograph, the snorkel is used to direct an elevated master stream into the classroom portion of the school building (center and left) in order to protect the gymnasium/ auditorium portion (to the right of the photograph). (Herb Spangler via Good Intent Hose Company, Llewellyn.)

Heavy fire erupts from the roof of the school as Minersville's snorkel works its stream from above. Snorkels at this time were revolutionizing firefighting in the United States, as they gave firefighters a platform from which to direct a large caliber stream with a great degree of accuracy and agility. (Herb Spangler via Good Intent Hose Company, Llewellyn.)

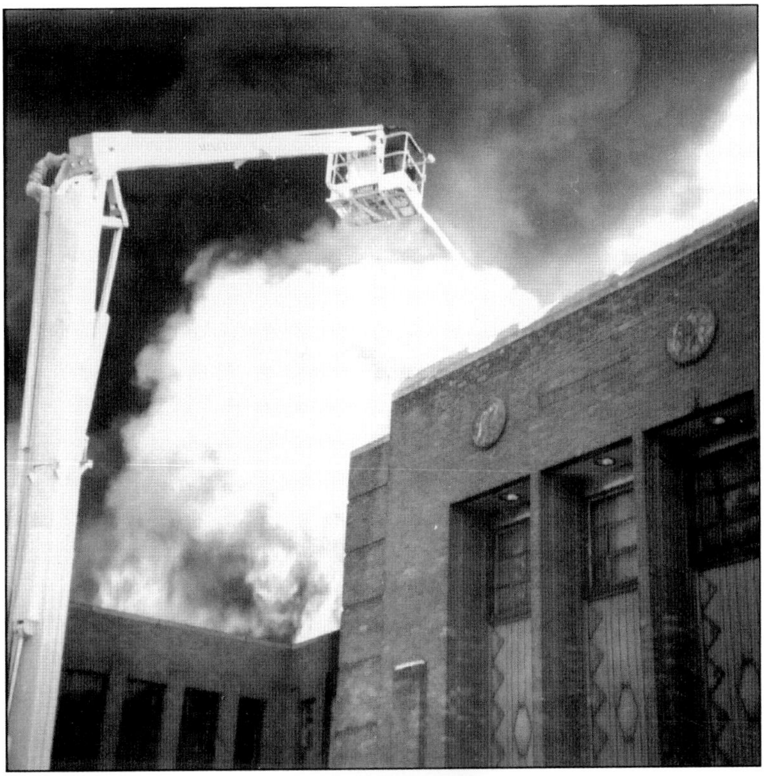

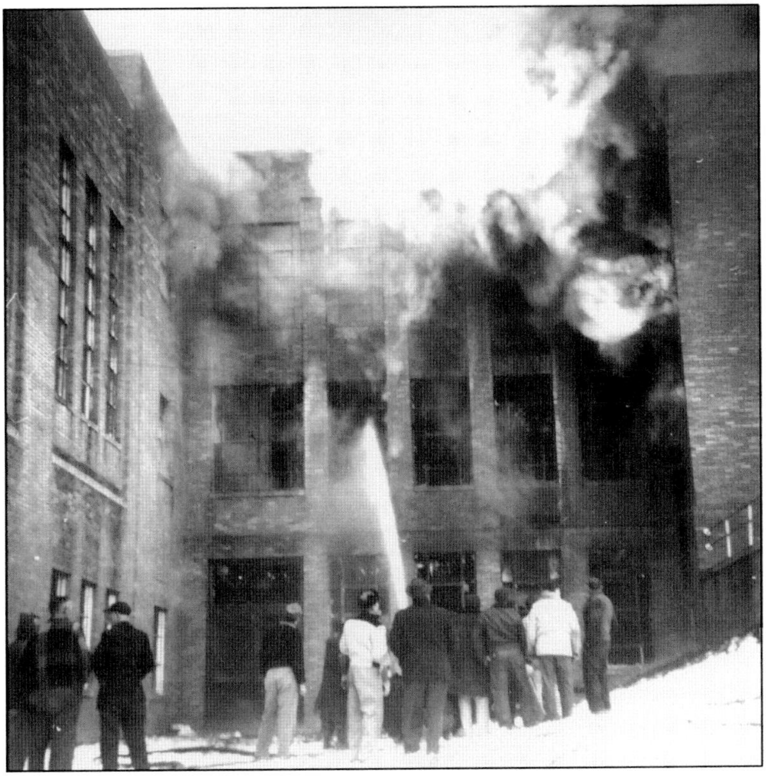

Firefighters direct a stream from a 2.5-inch hand line into the well-involved rear of the Llewellyn Elementary School. It was believed that the blaze originated in the basement area after a fresh fire had just been started in the furnace. Due to the slope of the terrain, the building rose to three stories in the rear and two stories in the front. (Herb Spangler via Good Intent Hose Company, Llewellyn.)

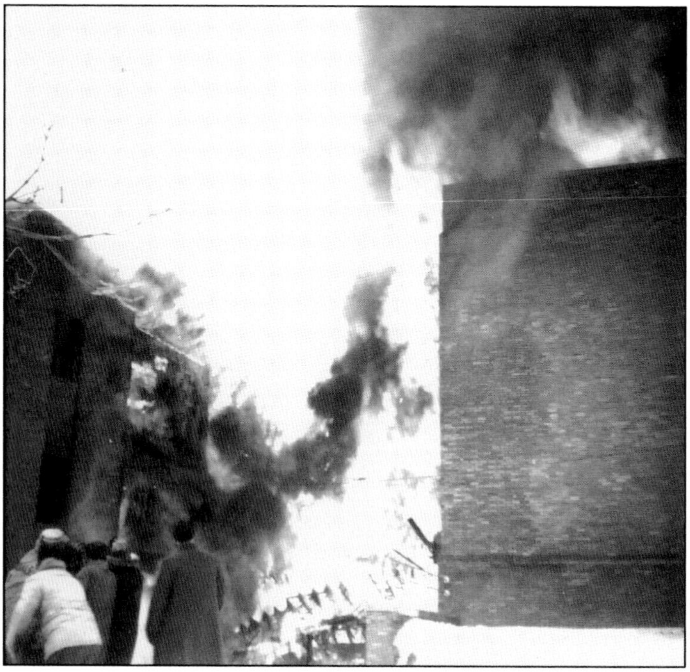

A major collapse had just occurred in the rear of the Llewellyn Elementary School when this photograph was taken. Thankfully no firefighters were injured. The gymnasium/auditorium was saved and still stands in virtually original condition. The fire loss was placed at over $600,000 in December 1963. (Herb Spangler via Good Intent Hose Company, Llewellyn.)

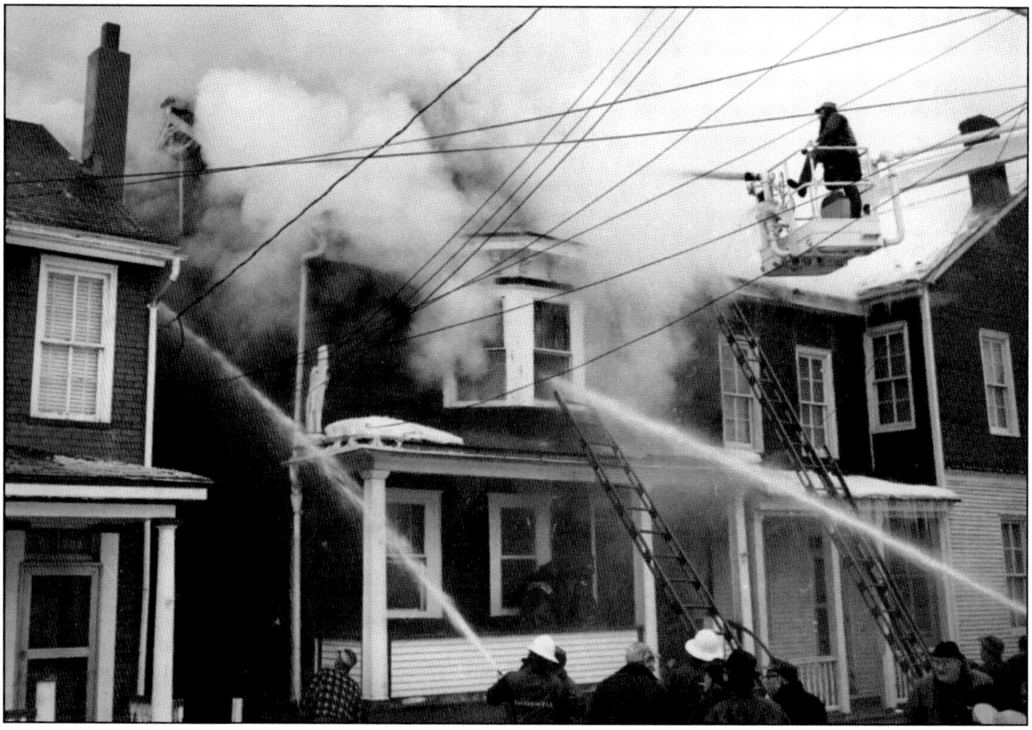

At 9:40 a.m. on a cold January 4, 1968, fire gutted this home at 306 Carbon Street in Minersville. The frame double-block home to the right on North Third Street was also heavily damaged. All five Minersville fire companies, as well as three Pottsville companies battled the blaze. Rescue Hook and Ladder of Minersville directs a stream into the attic of the home from the company's 1963 Ford/Pitman snorkel. (SHFS.)

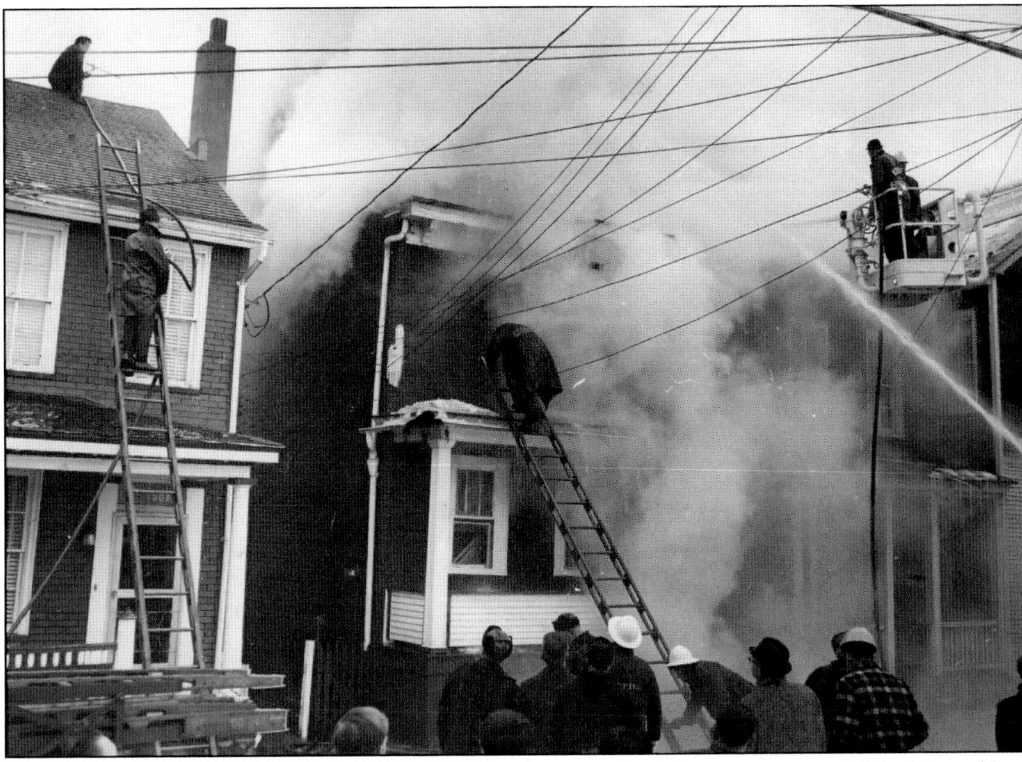

As the fire at 306 Carbon Street in Minersville advances, firefighters operate a 1.5-inch hand line from the platform of the Rescue Fire Company's snorkel. In addition, firefighters have stretched a booster line to the roof of the exposed home to the left of the original fire building. In the extreme lower left of the photograph is the rear of Rescue's 1941 American LaFrance city service ladder truck. (HSSC.)

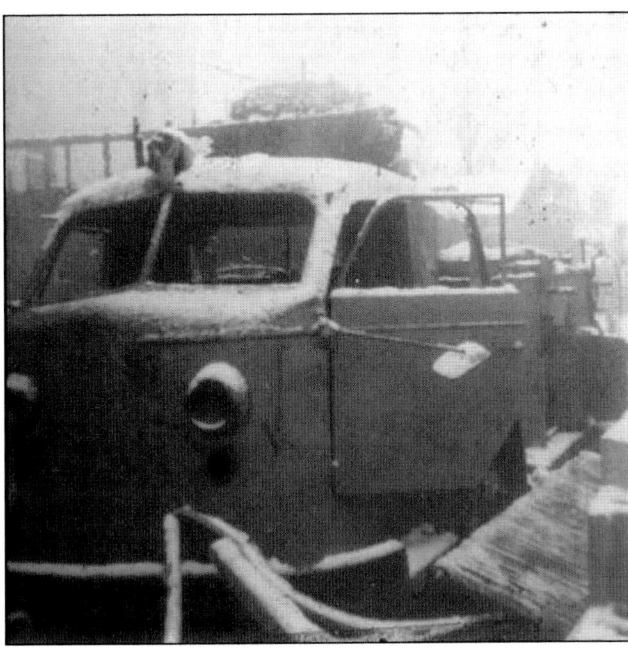

The 1956 American LaFrance 700 Series pumper of the Good Intent Hose Company, Llewellyn, was destroyed in a fire originating in the firehouse on February 5, 1970. A group of Girl Scouts boarding a bus in front of the firehouse discovered the fire. The alarm was sounded at 5:55 p.m. Thirteen fire companies responded to assist Llewellyn firefighters. (Good Intent Hose Company, Llewellyn.)

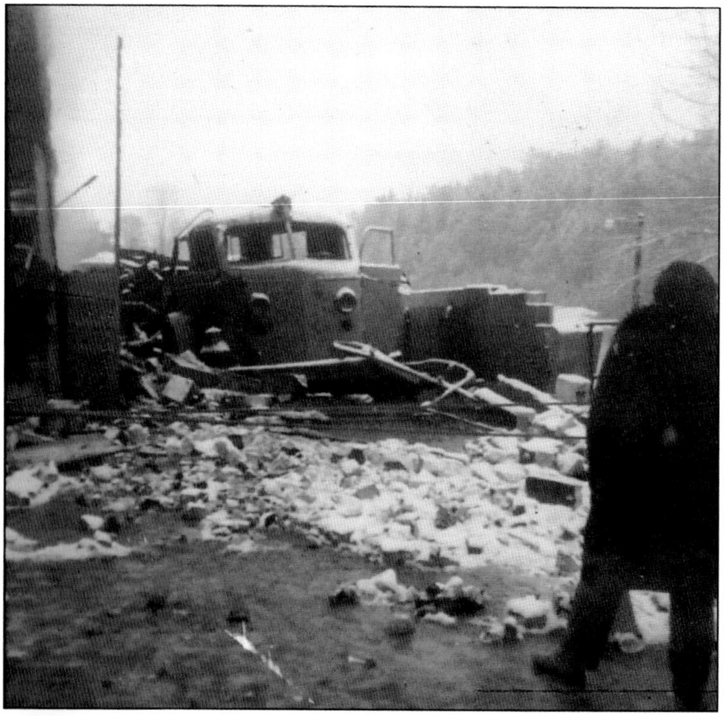

The Llewellyn firehouse was completely destroyed. The engine room was a block building attached to the original frame structure. Heavy fire conditions in the rear of the structure greeted arriving firefighters. The Reedy Gas Station on Route 209 (to the left of the firehouse in the photograph) was exposed by the fire and suffered minor damage. (Good Intent Hose Company, Llewellyn.)

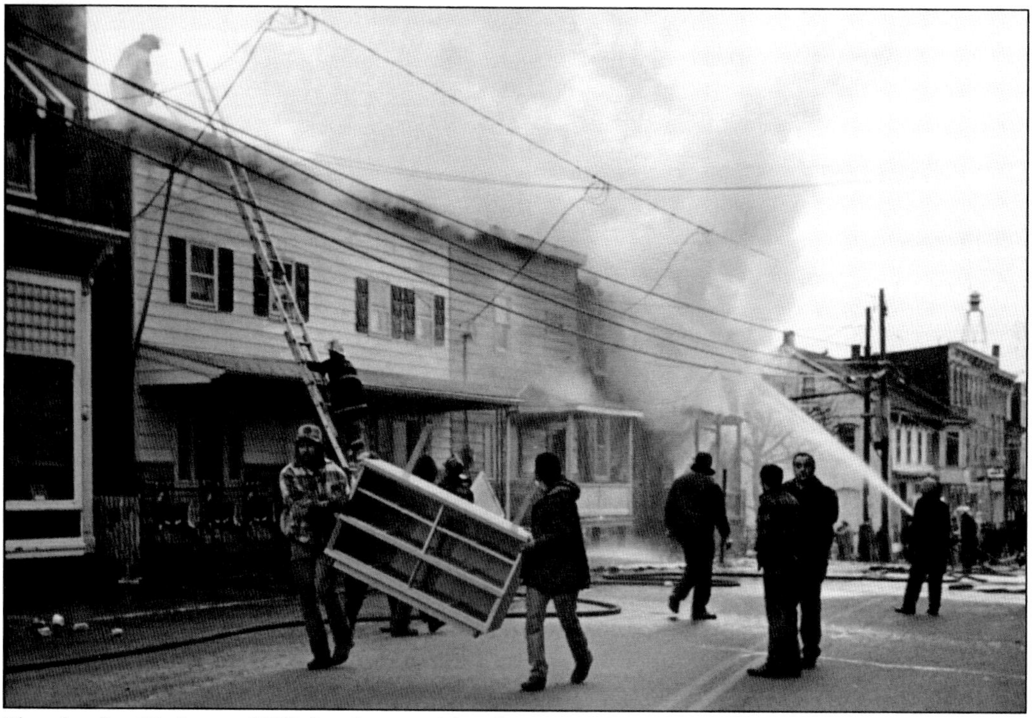

Shortly after Christmas 1977, fire destroyed four homes on East Main Street in Tremont. Firefighters responded to the late-afternoon alarm on December 27. As firemen work to access the roofs of the frame row houses, residents carry furniture saved from the flames. Fire companies from Tremont, Donaldson, Pine Grove, Ravine, Llewellyn, and Minersville responded. (HSSC.)

Firefighters and bystanders quickly retreat as the sidewall of this home collapses in the Black Heath section of Cass Township. The fire occurred on March 31, 1978. An occupant of the home smelled smoke at approximately 8:15 a.m. and called the fire department. The house was well involved when firefighters arrived. (HSSC.)

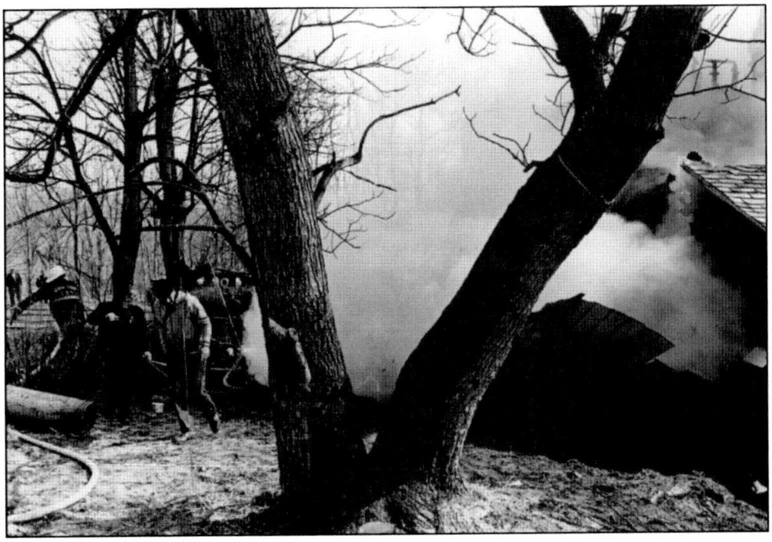

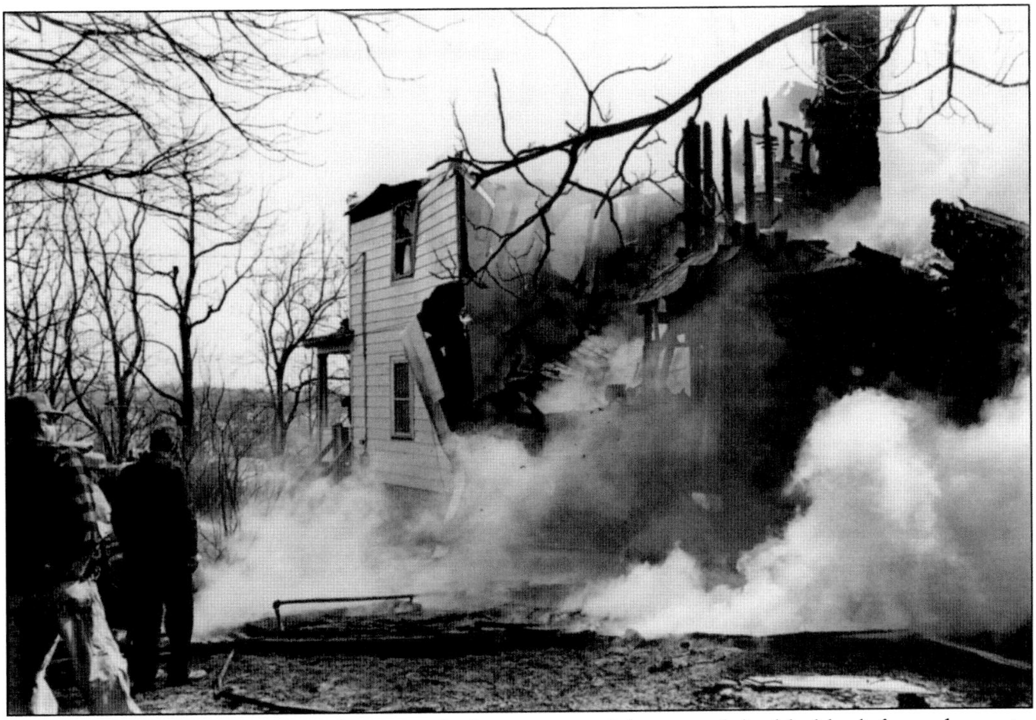

With much of the sidewall on the ground, the interior of the gutted double-block frame home in Black Heath, Cass Township is visible on March 31, 1978. The fire was under control by 10:00 a.m., but firefighters remained on the scene wetting down the remains throughout the day. Fire companies from South Cass, Forestville, and Minersville took part in the efforts. (HSSC.)

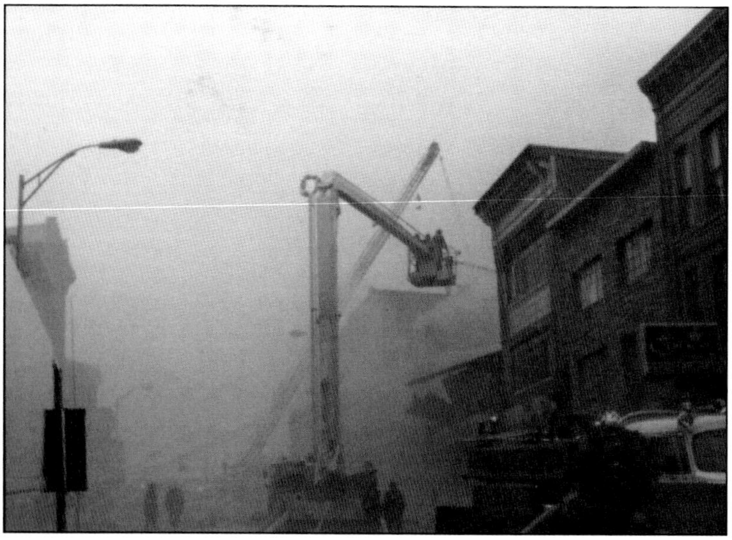

Daybreak reveals the destruction that fire caused in the 200 block of Sunbury Street in Minersville on November 25, 1979. Smoke still lingers in the street as Rescue Hook and Ladder's snorkel operates above the collapsed buildings. West End of Pottsville's aerial ladder is in the background. To the right is Mountaineer Hose Company's American LaFrance pumper. (Eric Eichenberg.)

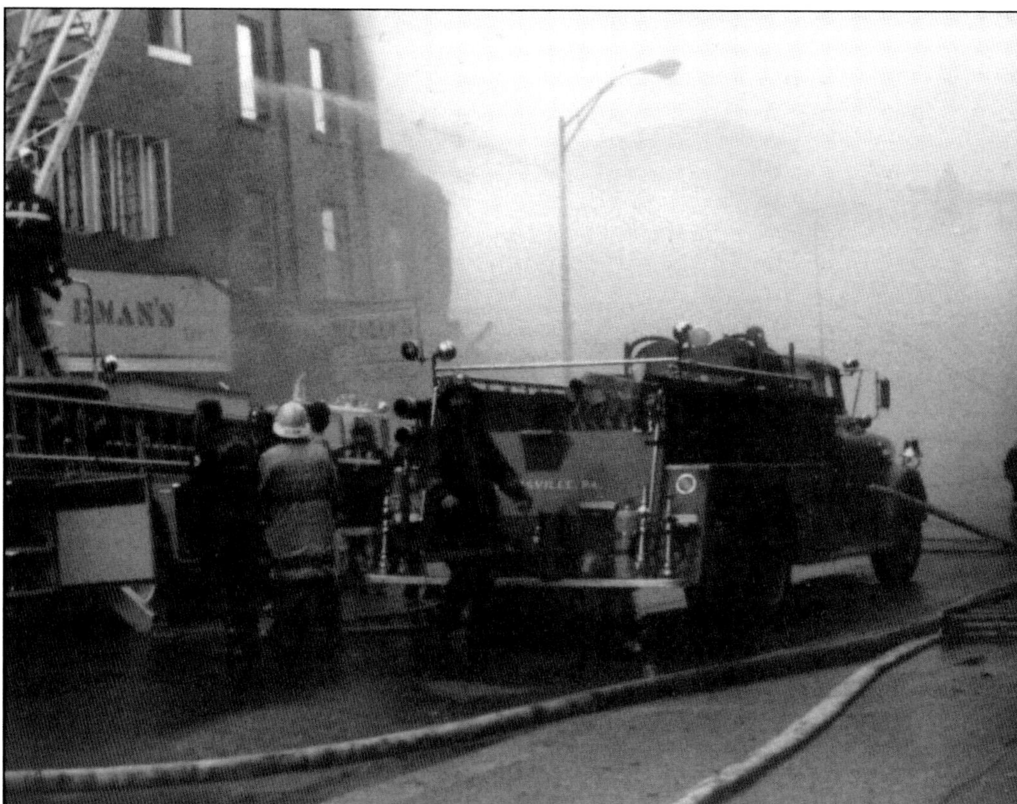

Looking East on Sunbury Street, the remains of one of the three buildings destroyed by fire is visible. Two other buildings were damaged, and 20 individuals were left homeless. The alarm was received at 3:50 a.m. Independent Hose Company of Minersville is in service with their 1958 GMC/Pirsch pumper while West End of Pottsville has the aerial ladder raised on their 1977 Seagrave quint. (Eric Eichenberg.)

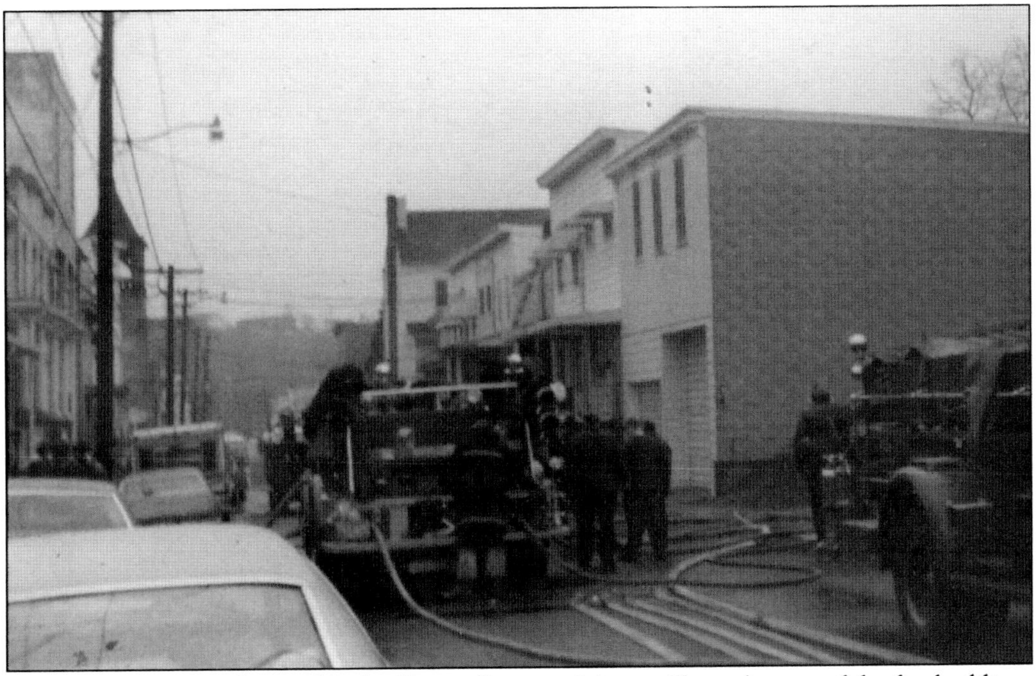

Fire units operate in the 200 block of Lewis Street in Minersville to the rear of the fire buildings on Sunbury Street on November 25, 1975. Hose lines are running from Good Will of Minersville's 1955 American LaFrance pumper. Rescue of Minersville's 1941 American LaFrance city service truck is visible to the right. The fire caused $400,000 in damage. (Eric Eichenberg.)

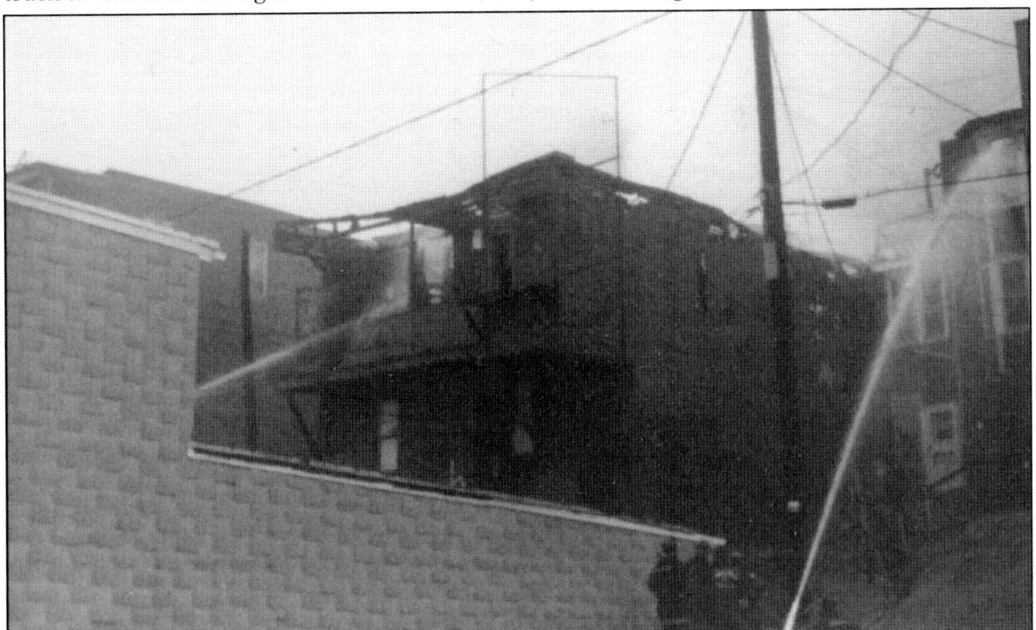

Heavy fire was showing from the rear of the buildings on Sunbury Street when firefighters arrived on the scene. This view is from Lewis Street in the rear as firefighters continue to pour water on the buildings. The cause of the fire was not determined. This block in Minersville was hit hard by fire again on February 13, 1992. (Eric Eichenberg.)

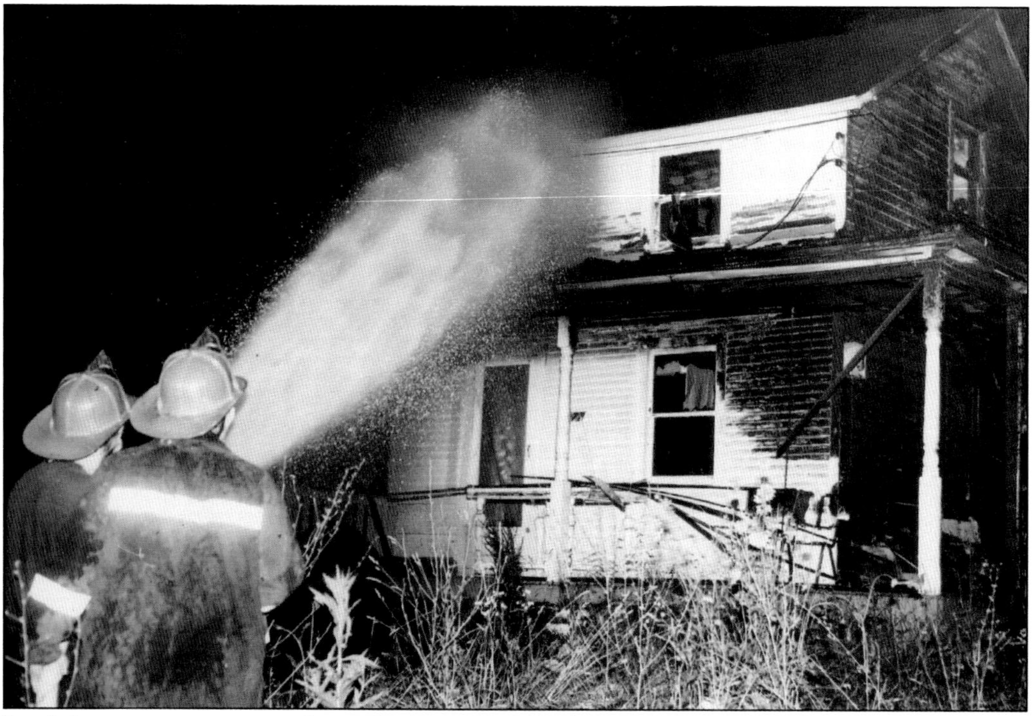

Firefighters continue to wet down the remains of this home in Cass Township that was destroyed on July 14, 1982. The early morning blaze also destroyed a garage. (HSSC.)

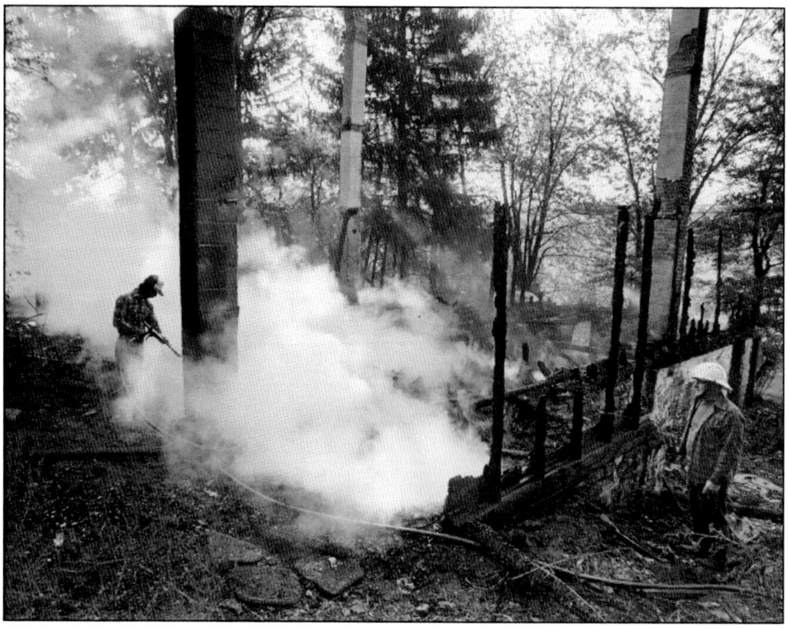

A firefighter from the Mt. Pleasant Hose Company uses a Hardie Gun–tipped booster line to soak the remains of this frame home that was destroyed on September 23, 1983. Firefighters arrived in response to the 1:00 a.m. alarm to find the vacant, two-story frame home fully involved. The home was located in Glen Carbon between Heckscherville and Buck Run. (HSSC.)

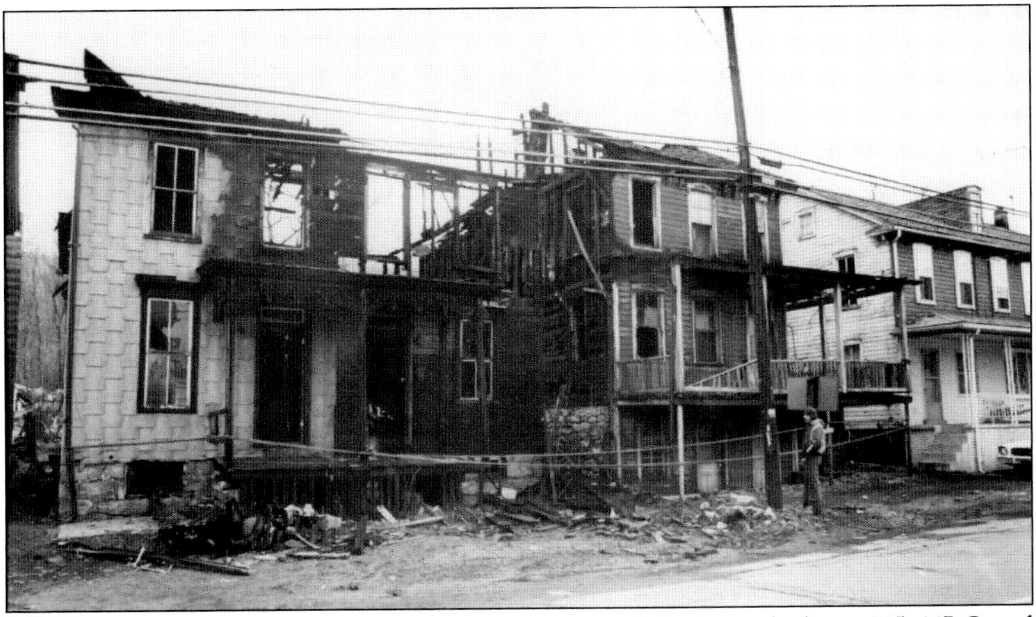

Fire in the Sheridan portion of Porter Township destroyed the frame duplex at 145–147 Grand Avenue, as well as the frame home at 143 Grand Avenue on March 20, 1982. Responding to the 10:30 a.m. alarm were the West End Fire Company, Sheridan, along with the Tower City, Reinerton, Wiconisco, and Williamstown fire companies. (HSSC.)

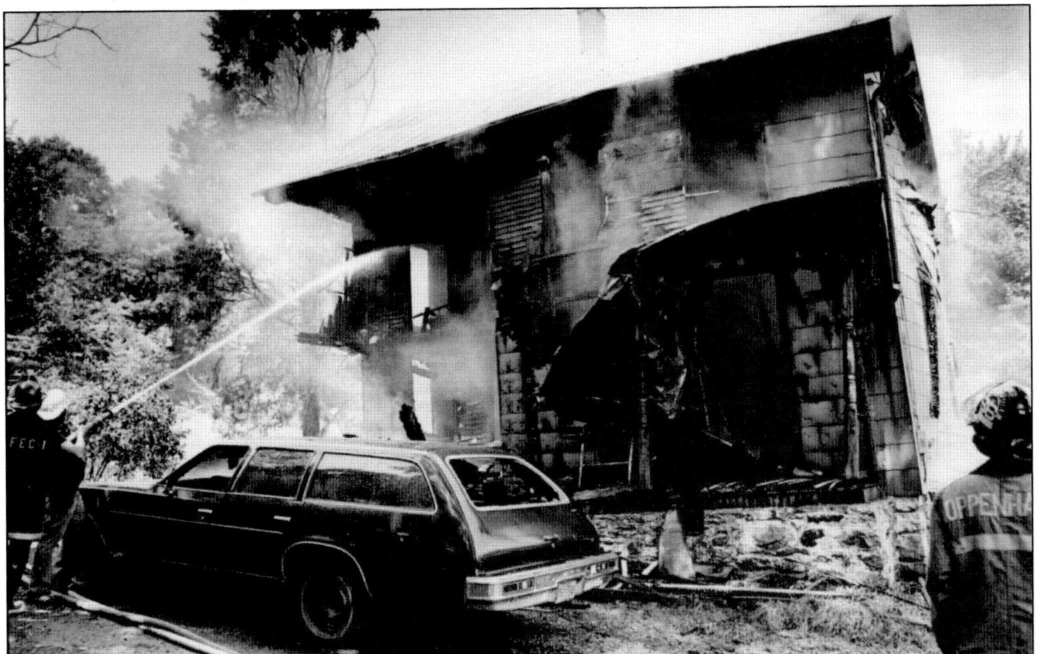

Firefighters continue to wet down the remains of this two-and-a-half-story frame home at 23 North Fifth Street in Tower City. Companies arrived on July 13, 1982, in response to a 1:45 p.m. alarm to find the home fully involved in fire. Also destroyed was the 1972 station wagon pictured here. Fire companies from Sheridan, Reinerton, Lykens, and Wiconisco assisted Tower City. (HSSC.)

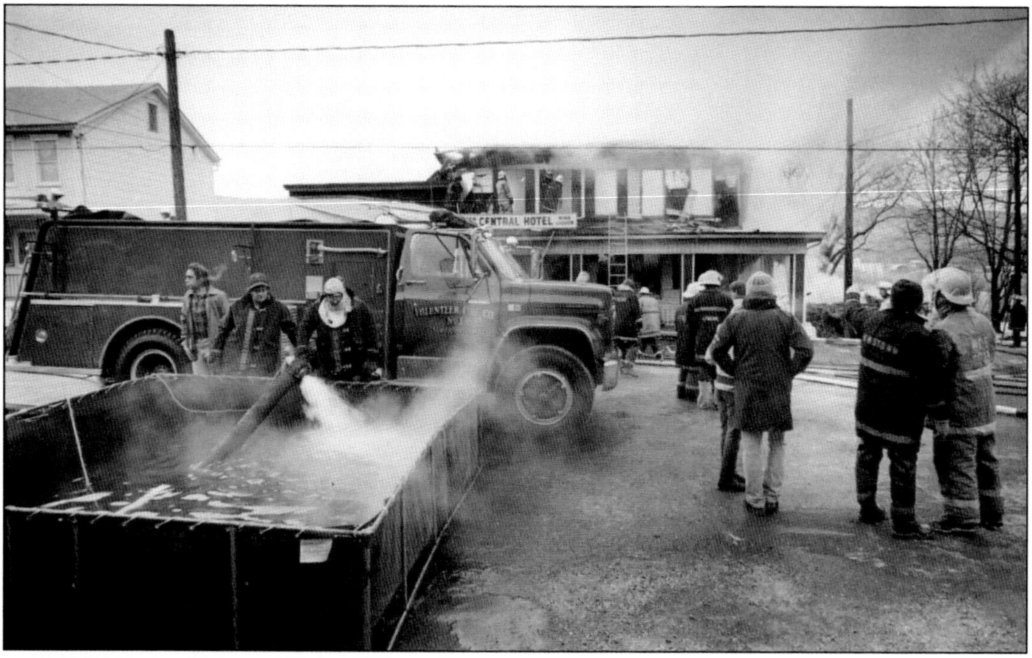

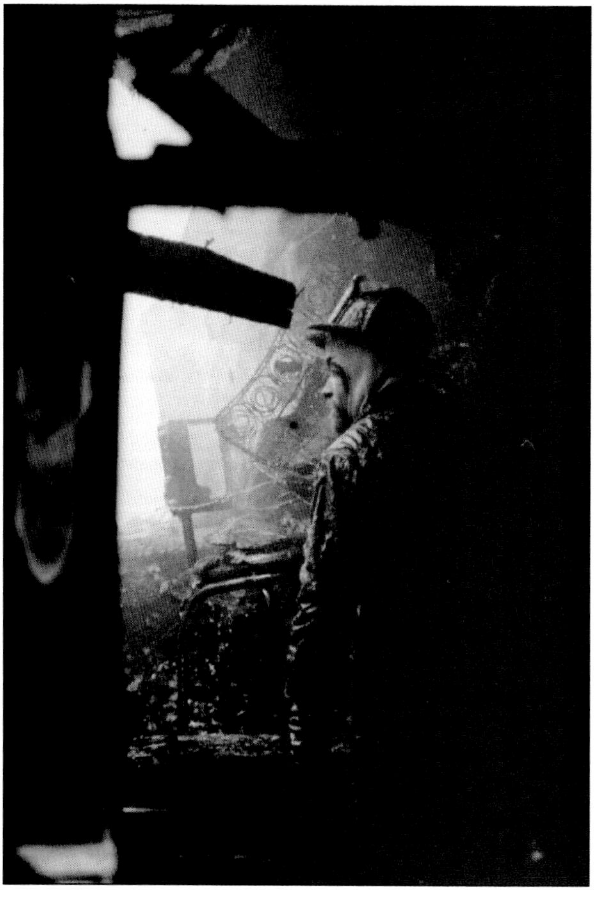

Tower City's Chevrolet pumper drafts water from a portable dam as daylight breaks over the remains of the Central Hotel in Orwin. Firefighters from the western end of Schuylkill County as well as upper Dauphin County fought the fire in the bitterly cold temperatures shortly after Christmas 1983. (HSSC.)

An ice-coated firefighter surveys the gutted first floor of the Central Hotel in Orwin. The alarm was sounded at 3:40 a.m. on December 27, 1983. It took three hours to bring the fire under control. (HSSC.)

Firefighters from Minersville extinguish a flammable-liquids fire as part of their fire-training class on February 19, 1973. The classroom portion of the program was held at the Good Will Fire Company, Minersville, while the practical exercises were held at the old dump site off Delaware Avenue. (HSSC.)

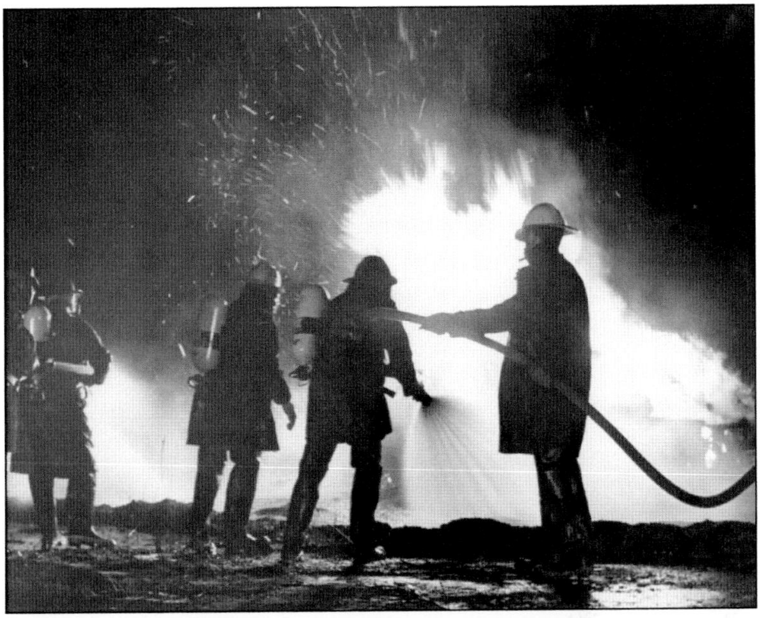

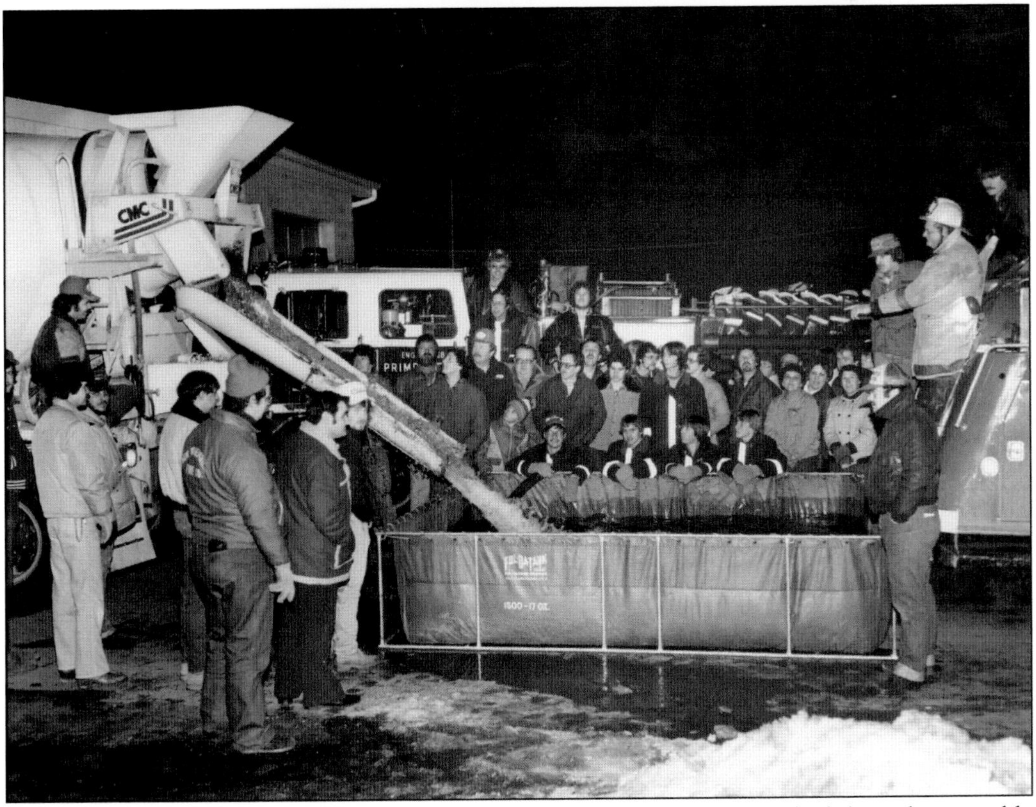

South Cass Fire Company of Primrose has a 1978 Hahn pumper set up to draft from this portable dam, which is being filled with water by a cement mixer hauling 3,000 gallons of water. State fire instructor Thomas Towers of Ashland taught the class in which the creative use of available water resources was demonstrated on February 1, 1982. (HSSC.)

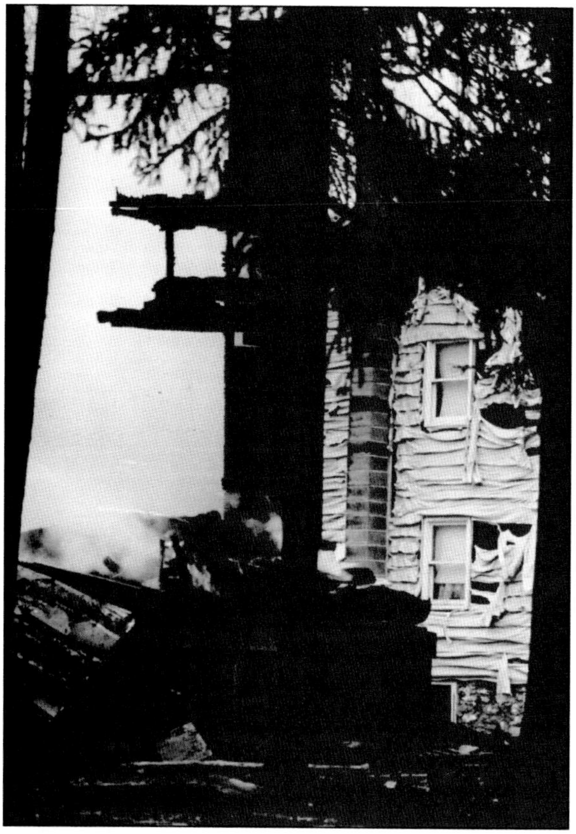

The melted siding on the home in the background is testimony to the heavy fire conditions that firefighters encountered when they arrived at this home in Newtown on December 28, 1982. The remains of the original fire building smolder in the foreground. (HSSC.)

Firefighters continue to hose down the remains of the two-story frame home on Pine Street in Newtown on December 28, 1982. The alarm was sounded at 6:50 a.m. Fire companies from Pine Grove, Branchdale, Tremont, Donaldson, Ravine, Valley View, and Phoenix Park assisted Newtown firefighters. (HSSC.)

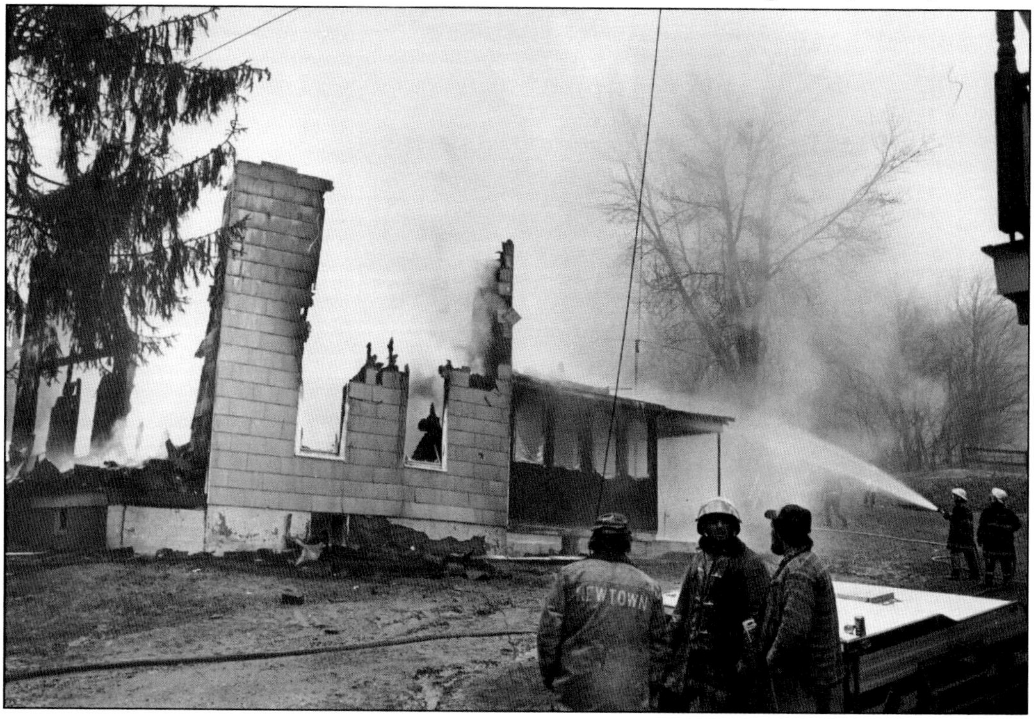

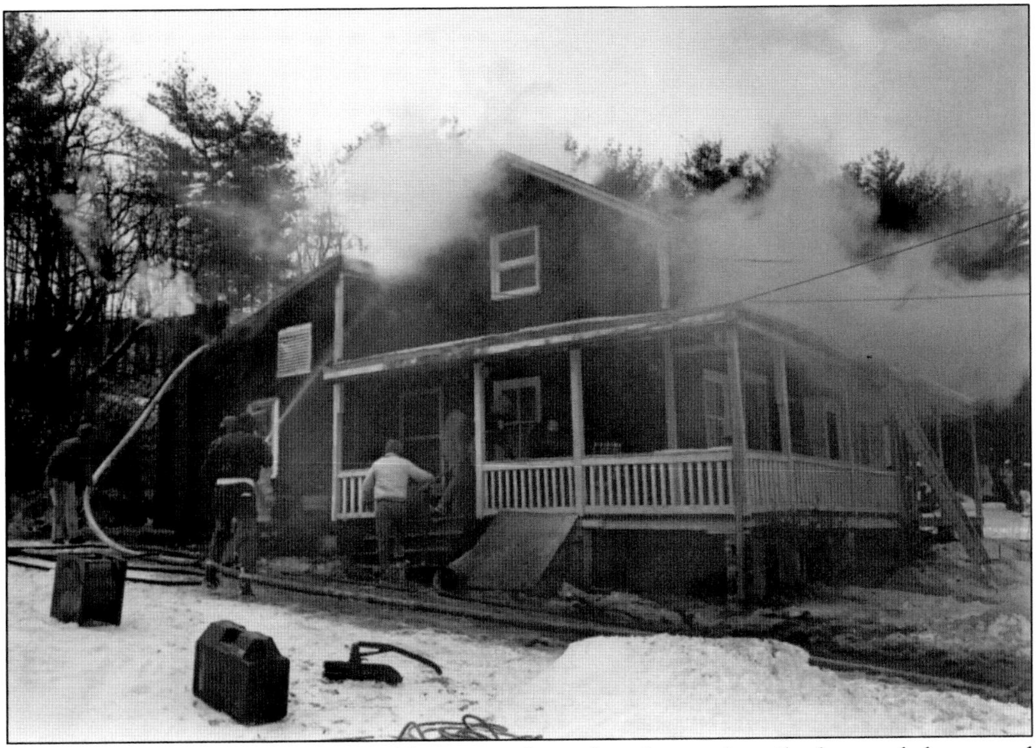

A fire believed to have been caused by an overheated coal stove heavily damaged the second floor and attic of this frame home in the Greenfield section of Forestville on January 15, 1982. Firefighting equipment rests in the yard: smoke ejector, SCBA case, and hose clamp. (HSSC.)

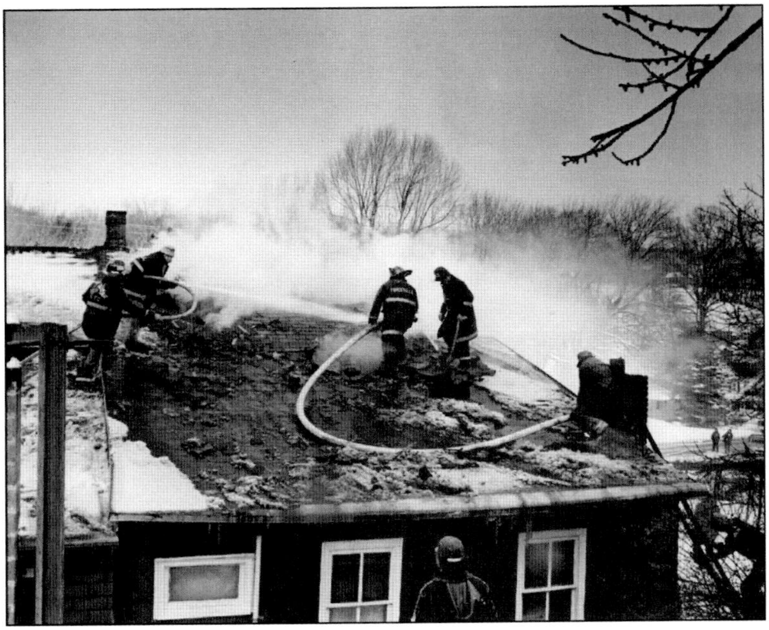

Firefighters maneuver to expose fire in the attic area of the home in Forestville on January 15, 1982, while cautiously working on the snow and ice accumulation on the roof. The home was located between Forestville and Branchdale. The alarm was received at 10:30 a.m. (HSSC.)

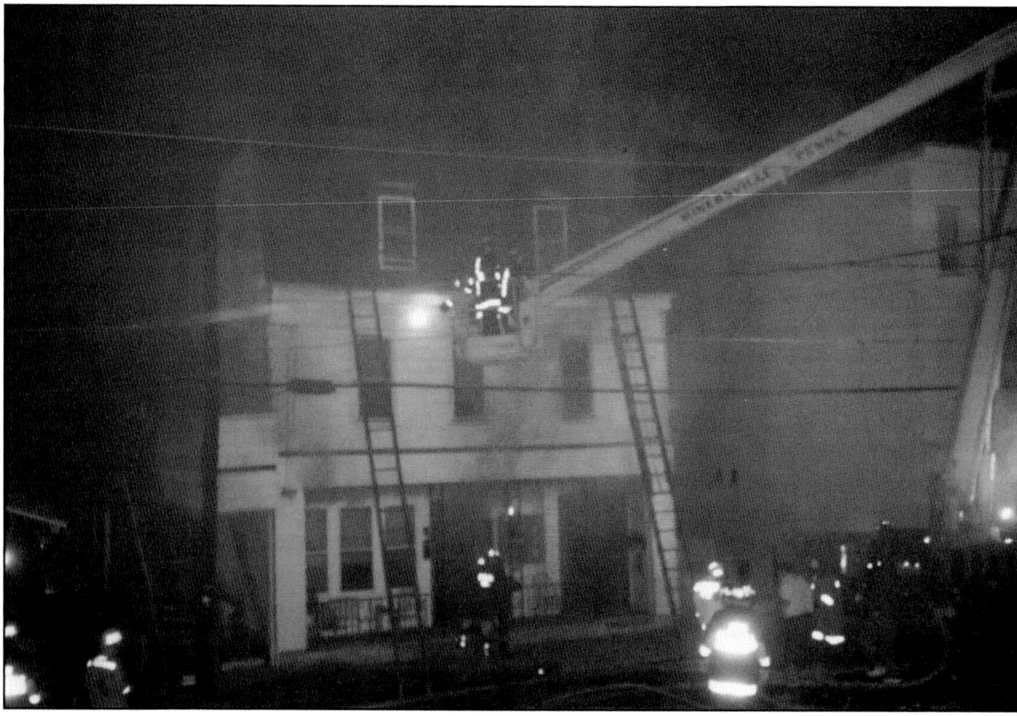

This hazy photograph is the result of smoke pushing from this three-story frame apartment building on New Castle Street in Minersville. Rescue Hook and Ladder is "booming-up" with its snorkel as hose lines are stretched. The fire occurred on a balmy morning in August 1990. (Authors' collection.)

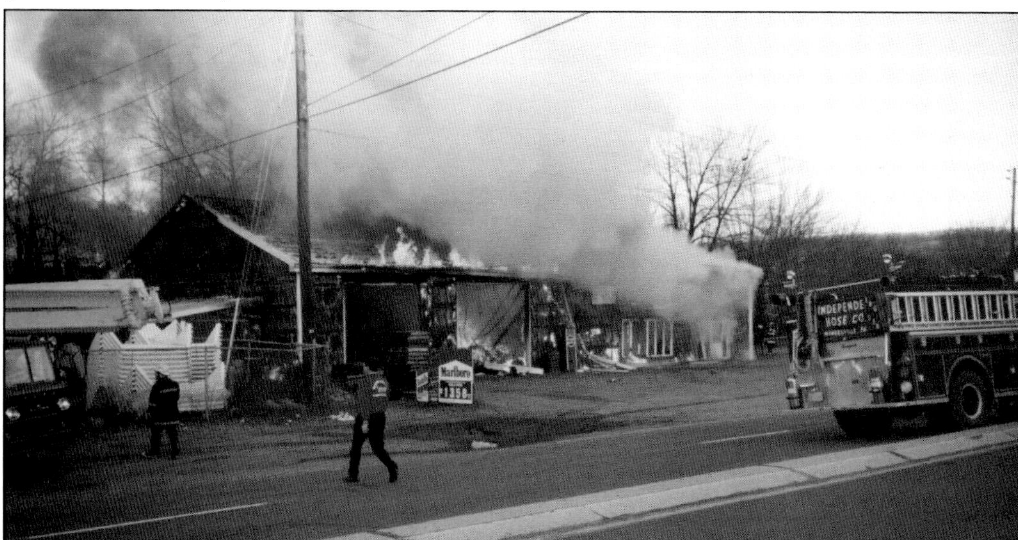

This beverage distributorship was completely involved in fire when companies arrived in response to the mid-afternoon alarm in December 1990. Minersville's snorkel (left) backs into position to protect a commercial building exposed by the fire as Independent's 1987 Seagrave pumper arrives on the scene (right). The business was located on the Pottsville/Minersville Highway. (Authors' collection.)

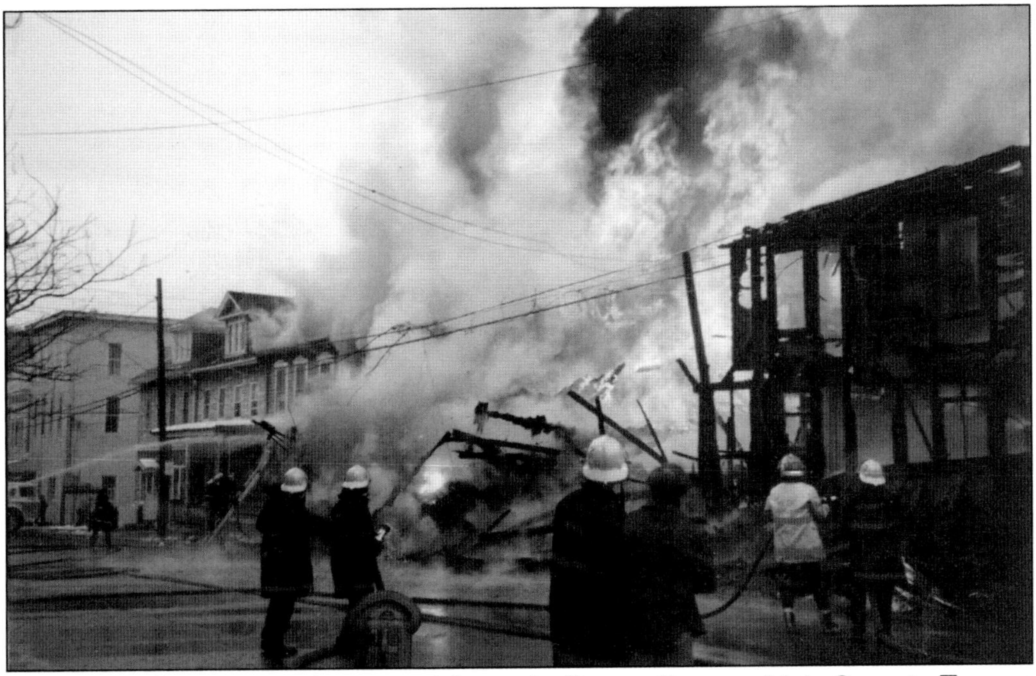

Firefighters all turn to watch as several frame dwellings collapse on Main Street in Tremont in January 1987. Members of companies from throughout the southern and western sections of Schuylkill County assisted Tremont firefighters in battling the mid-afternoon blaze, which destroyed seven buildings. (Authors' collection.)

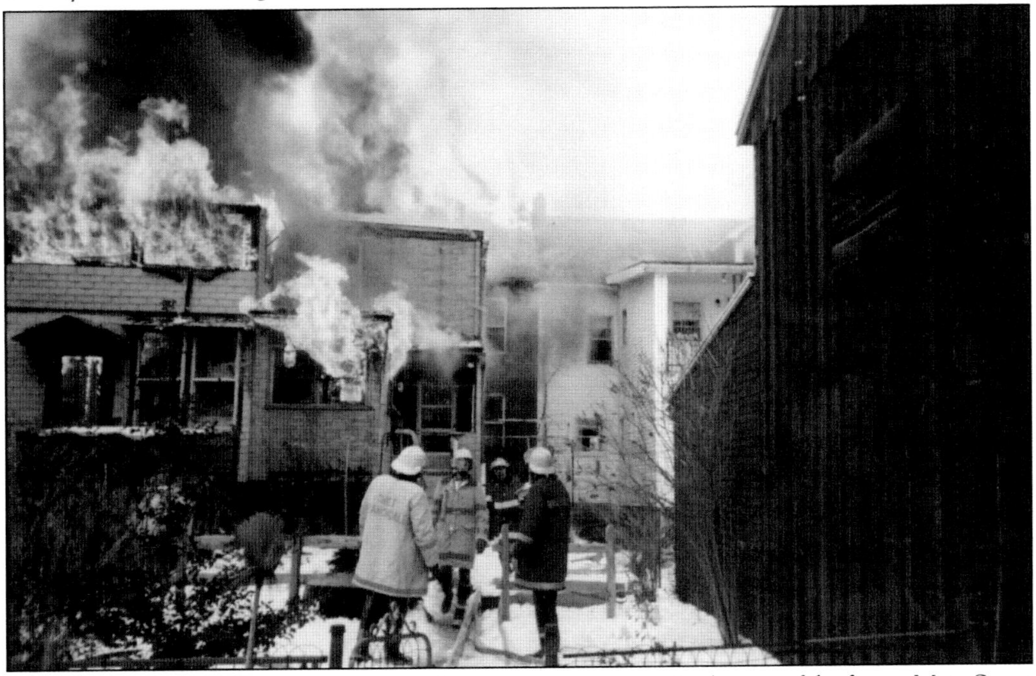

Firefighters from Branchdale attempt to get hose lines into position in the rear of the fire on Main Street in Tremont. Water supply issues hampered firefighters for a time. In the photograph, heavy fire consumes these frame dwellings and threatens additional properties in the block. (Authors' collection.)

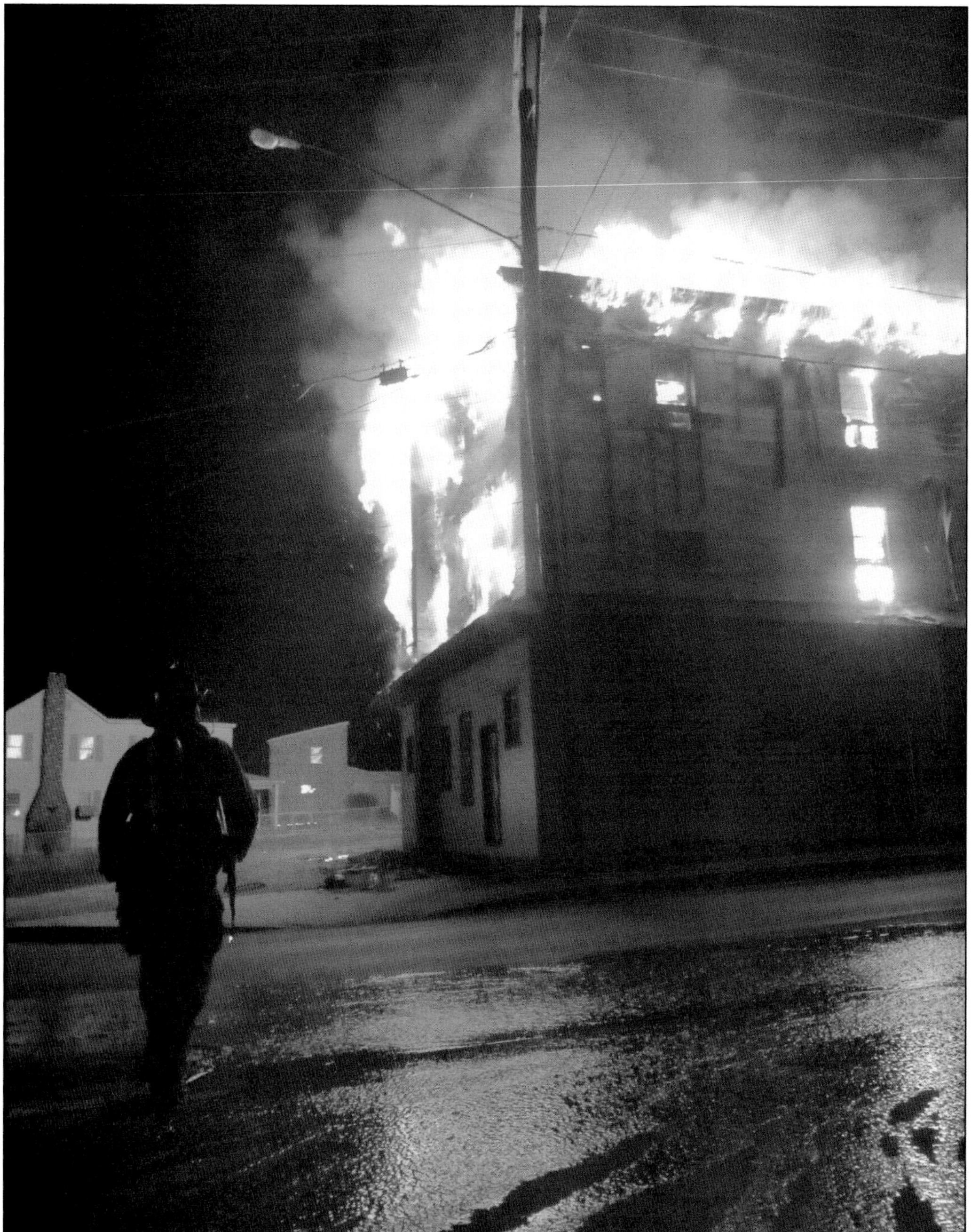

The 100-year-old Llewellyn Hotel was destroyed by fire on March 10, 2009. The alarm was transmitted at 12:39 a.m. The three-story frame building located at Bunting Street (Route 209) and Phoenix Park Road was vacant at the time of the fire. Fire companies from around the Llewellyn area assisted. Water was trucked in using fire department tankers to supplement the municipal water supply. Firefighters managed to prevent the fire from extending to an occupied frame dwelling just a few feet from the hotel. The fire was ruled as an arson. (Good Intent Hose Company, Llewellyn.)

Five

Central

With the discovery of anthracite nearby in 1790, Pottsville and its surrounding communities soon benefitted from an economic onslaught. From a sleepy settlement in the early 1800s, Pottsville quickly grew to a city with a population of over 30,000 by the early 20th century. With this rapid development came the need for fire protection. Pottsville organized its Humane and Schuylkill Hydraulians (Phoenix) fire companies in 1829, followed by the Good Intent and Rough and Ready (American Hose) in 1846 and 1848, respectively.

With further growth, new city fire companies emerged, including Atkins, West End, and Good Will. A western expansion of Pottsville soon brought the Yorkville Hose Company into the city's fire department ranks. Two interesting notes are that several of Pottsville's fire companies organized junior fire companies (boys' companies) during the mid-19th century; and, with the threatened destruction to the nation's capital in 1861, many Pottsville firefighters immediately enlisted and became known as the nation's "First Defenders" during the Civil War.

The neighboring boroughs of Port Carbon and Mount Carbon along the Schuylkill Canal formed their fire companies in 1879 and 1910. Mechanicsville Borough, just east of Pottsville, organized its fire protection in 1903. East Norwegian Township created three fire brigades in the villages of Diener's Hill, Schoentown, and Mill Creek beginning in 1915. Later in the 20th century, Norwegian Township started fire companies in the villages of Mar Lin and Seltzer in 1932 and 1945, respectively.

Just north of Pottsville, the coal-mining borough of St. Clair initiated five fire companies beginning in the 1860s. The first, the Columbia Hose Company, commenced in 1868. In addition to the Columbia, St. Clair added the Alert Fire Company in 1883, the Rescue Hook and Ladder in 1898, and the Phoenix of East Mines in 1911. St. Clair's youngest fire company, the West End of Arnots Addition, formed in 1951.

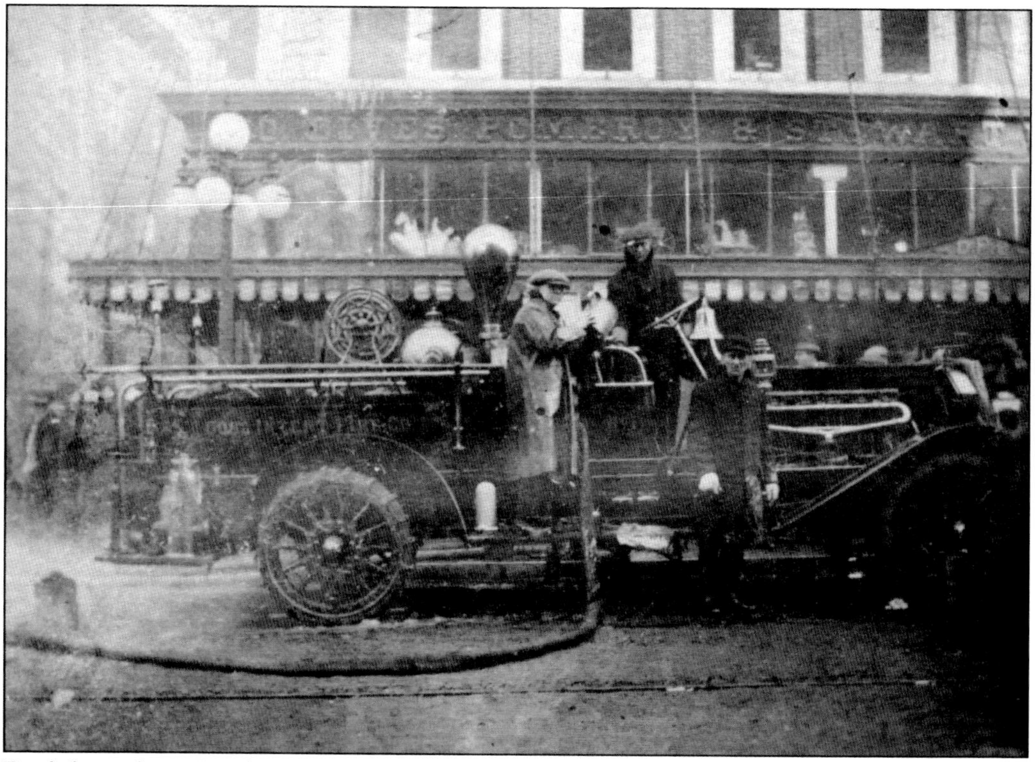

Firefighters from Good Intent, Pottsville, tend to their 1911 Robinson 700-gallons-per-minute pumper at the devastating Woolworth Fire, which destroyed the entire block along Centre Street from Norwegian to Mahantongo to Second Streets in Pottsville on December 17, 1914. The engine worked continuously for 13.5 hours that day. Behind the rig is Pomeroy's Department Store. (HSSC.)

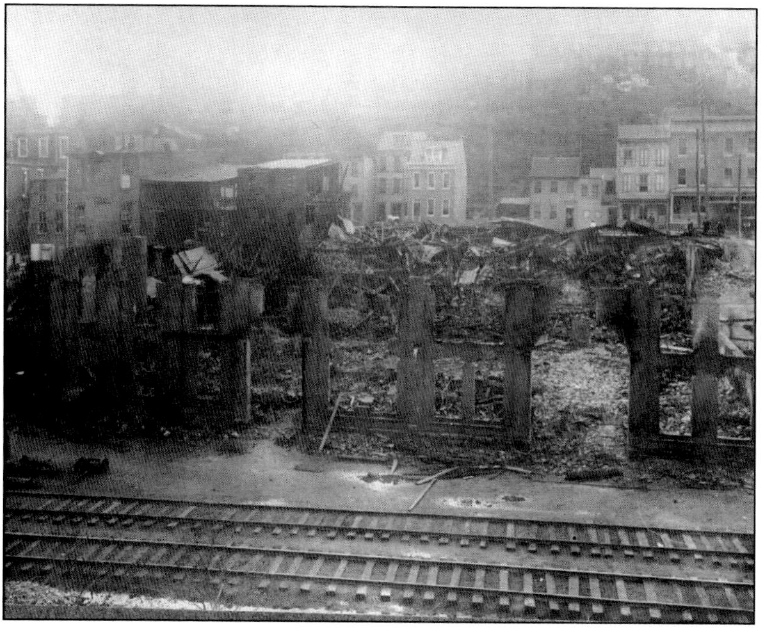

On January 6, 1917 (the height of World War I), an alarm was received for the Eastern Pennsylvania Railway Company car barn and electric-generating facilities in Palo Alto at 6:10 p.m. The company operated the trolleys that were familiar sights on local streets in the early part of the 20th century. This view shows the rear of the car barn and the homes along Bacon Street in Palo Alto. (HSSC.)

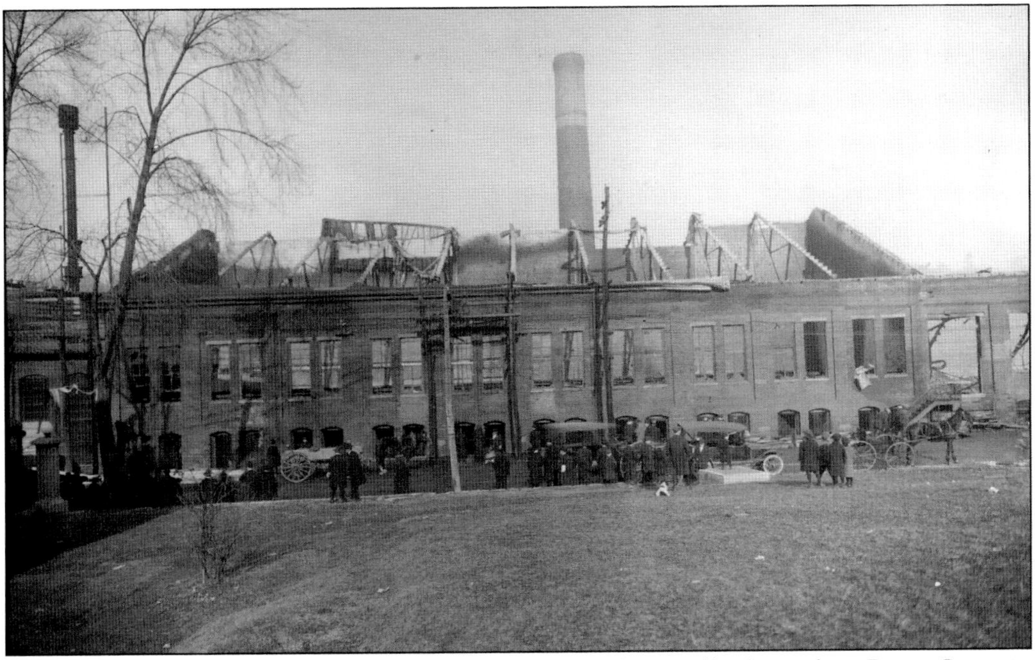

Spectators view the remains of the massive East Penn Railway trolley barn along Bacon Street in Palo Alto. A horse-drawn carriage and two automobiles reflect the changing modes of transportation at the time. (HSSC.)

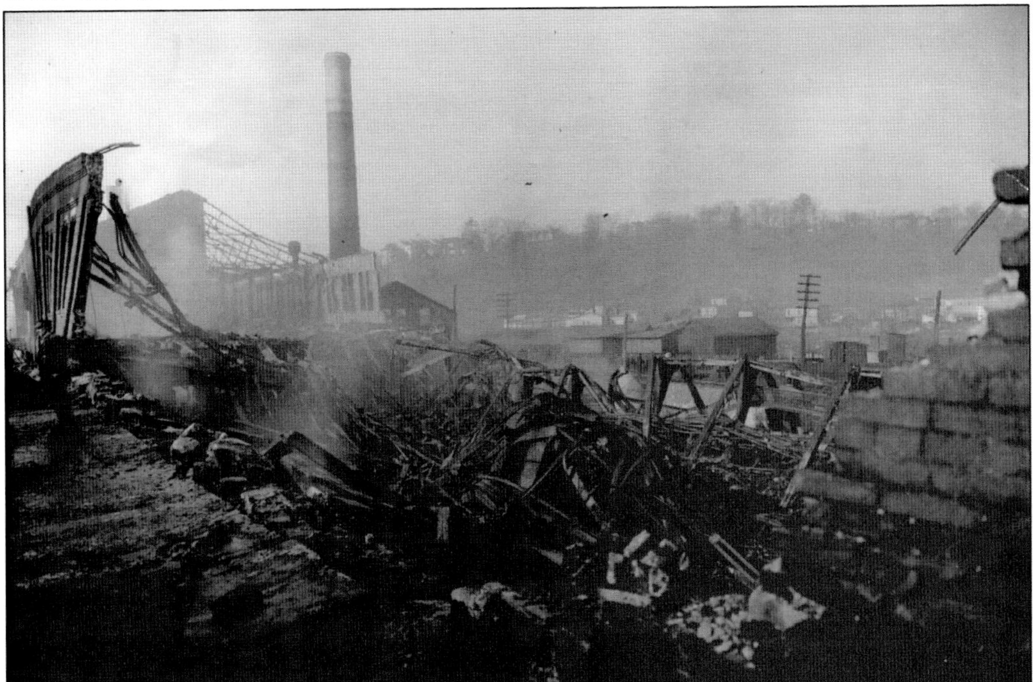

This photograph, also taken along Bacon Street in Palo Alto, looks down into the destroyed trolley barn. This gives some perspective of the massive size of this building and the firefighting challenges it posed in 1917. (HSSC.)

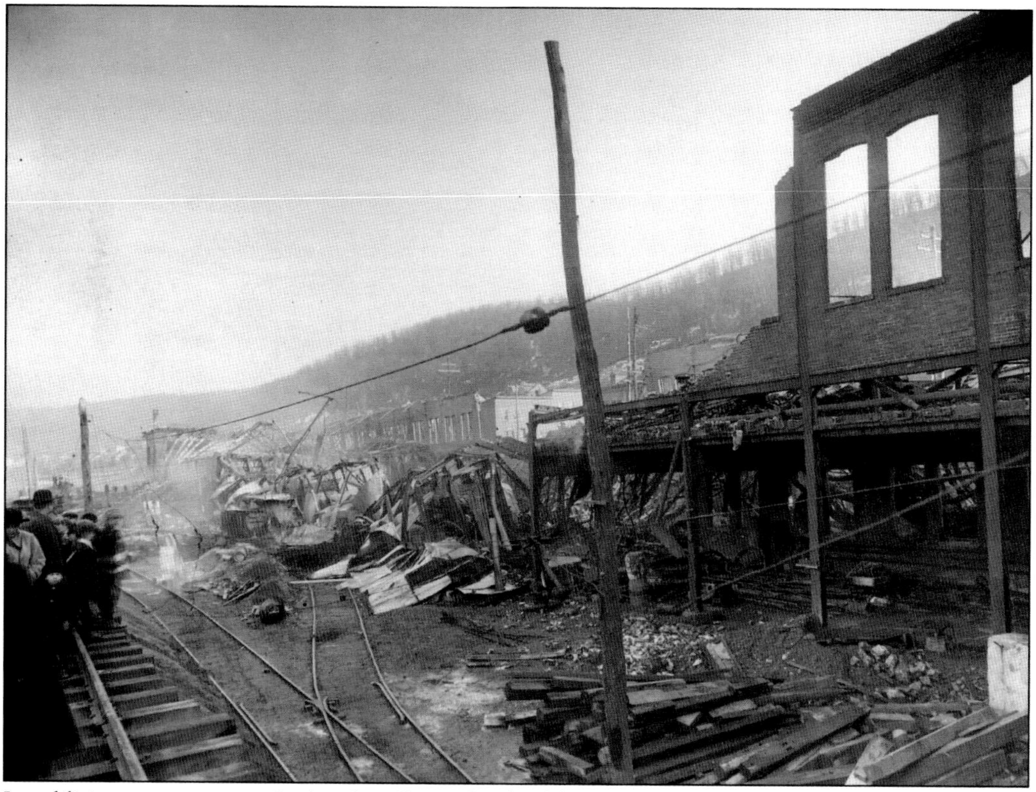

In addition to powering the local trolleys, this building generated electricity for a large portion of the surrounding area. This fire caused the loss of power to several thousand local residents as well as the Pottsville Hospital. Trolley tracks can be seen disappearing into the destroyed building. (HSSC.)

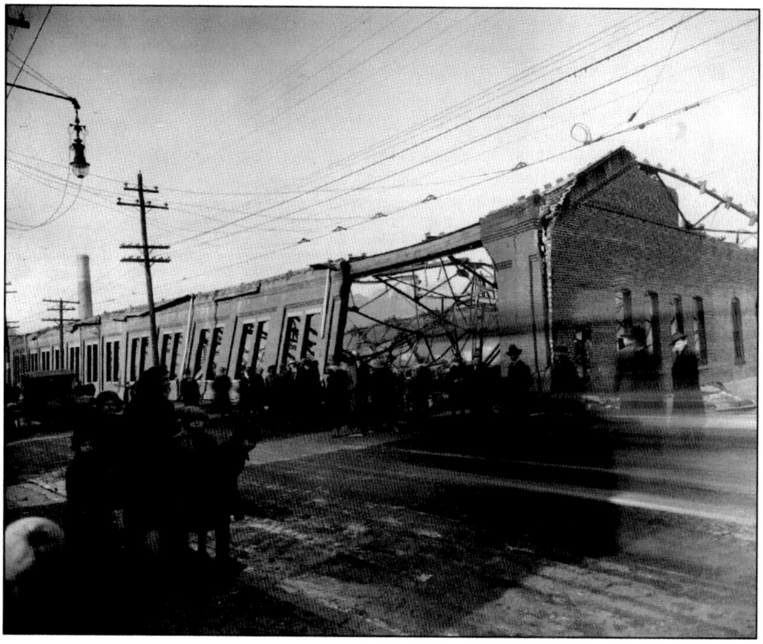

A mother and her children had evidently come to observe the destruction of the Eastern Pennsylvania Railway Company following the January 6, 1917, fire. The children, as well as several of the men across the street, seem more interested in the photographer than the remains of the fire building. (HSSC.)

The large generators used by the Eastern Pennsylvania Railway Company to power its electric trolleys, as well as surrounding homes and businesses, are shown in the fire-ravaged building. Fire companies from Palo Alto, Port Carbon, Pottsville, and Saint Clair fought the blaze. (HSSC.)

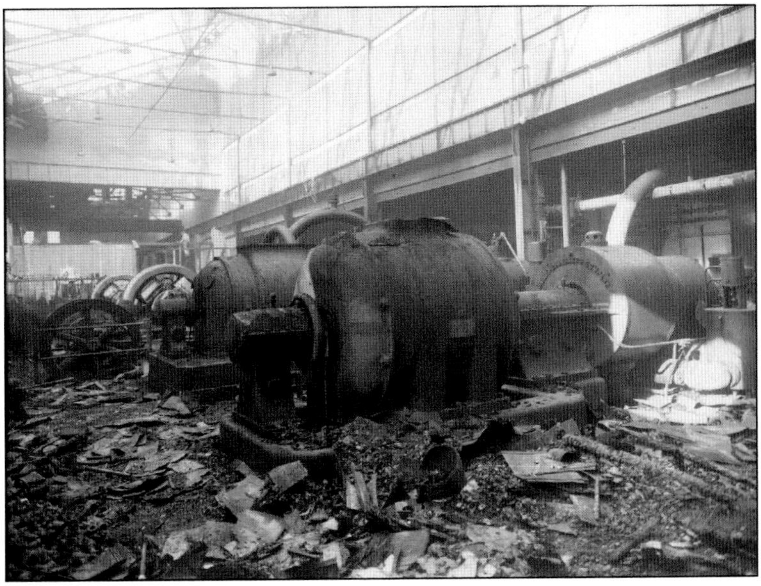

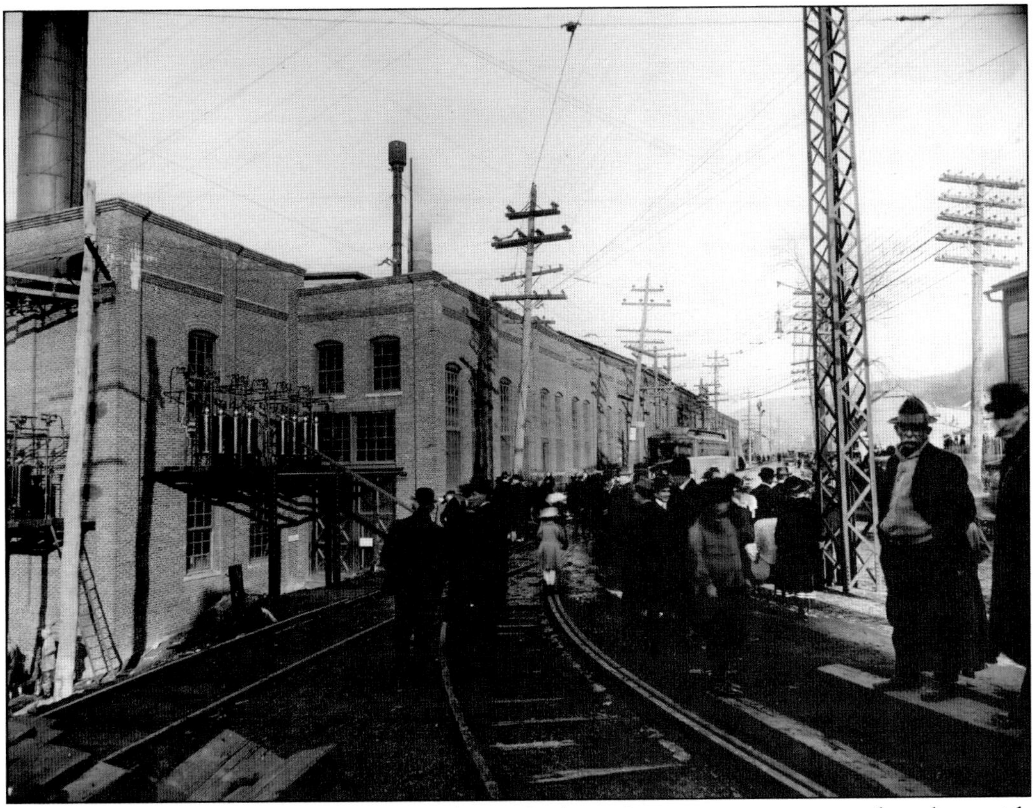

At the western end of the Eastern Pennsylvania Railway Company, spectators gather along with one of the trolleys supplied by the company. Firefighters managed to reduce the amount of damage at this end of the building. Interestingly, the frame warehouse barely visible in the extreme right of this photograph was destroyed in a spectacular fire in 1989. (HSSC.)

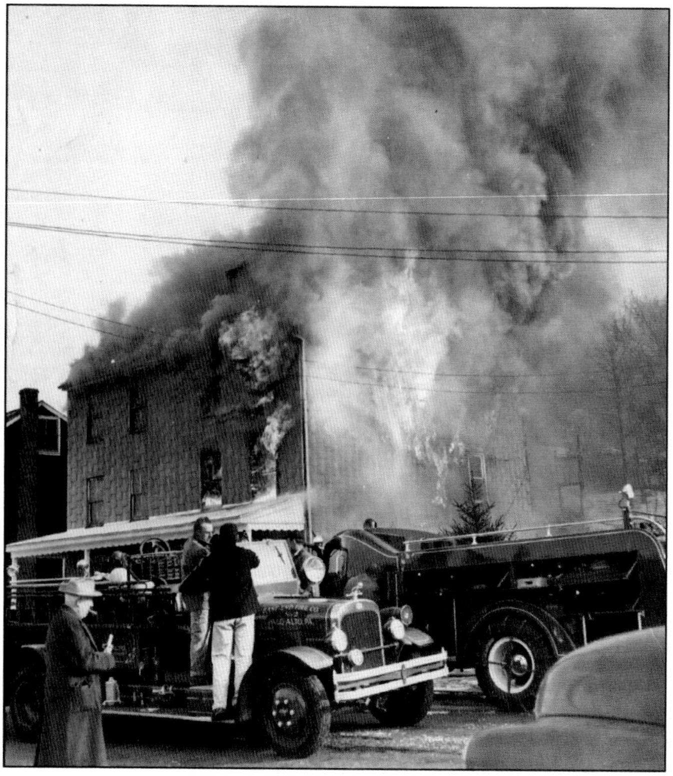

On a cold January 14, 1957, day at approximately 2:15 p.m., firefighters from Palo Alto arrived to find this wood duplex on Bacon Street well involved in fire. The 1931 Seagrave 500-gallon pumper of the East End Fire Company, Palo Alto, is in the foreground, and the 1952 Mack 500-gallon pumper of the Citizens Fire Company, Palo Alto, is in the background. (SHFS.)

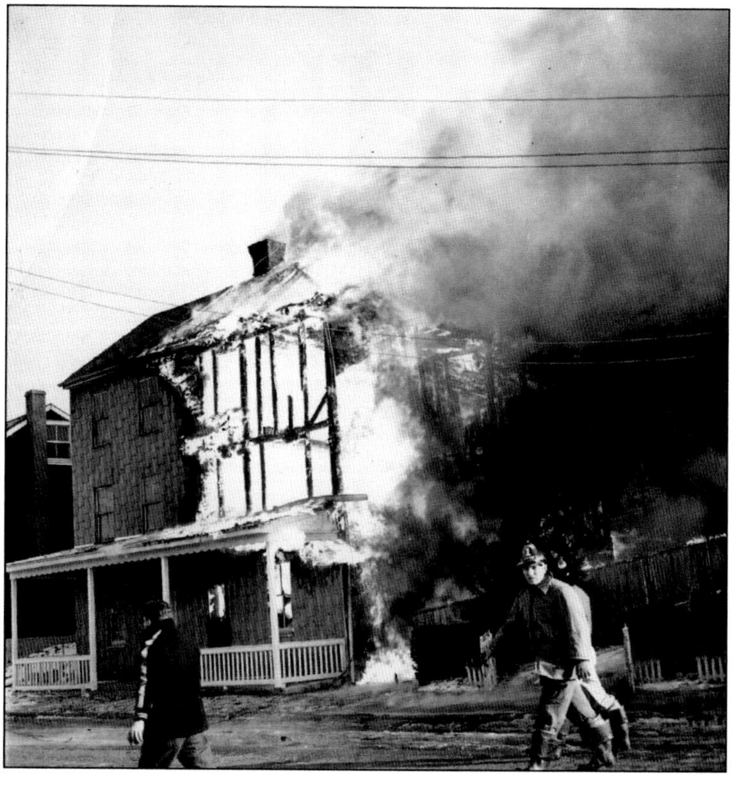

Water-pressure issues forced the firefighters and fire apparatus to retreat as fire consumed the homes in Palo Alto. The houses on either side of the duplex were also heavily damaged by fire. Pottsville sent fire apparatus from the American Hose, Greenwood Hill, and Yorkville fire companies to assist the Palo Alto Fire Department. (SHFS.)

Two months later, on March 14, 1957, Palo Alto firefighters were again faced with a serious fire on Bacon Street. At approximately 7:45 p.m., flames heavily damaged this frame duplex. Pottsville again sent the Greenwood Hill, American Hose, and Yorkville fire companies to assist the Palo Alto Fire Department. Schuylkill Haven, Cumbola, and Port Carbon fire companies were also on the scene. (SHFS.)

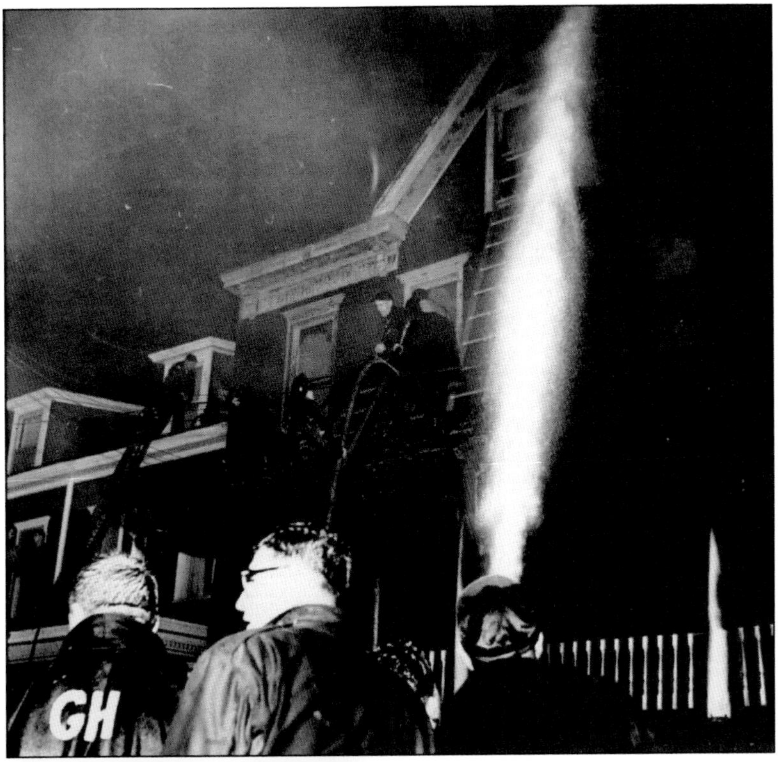

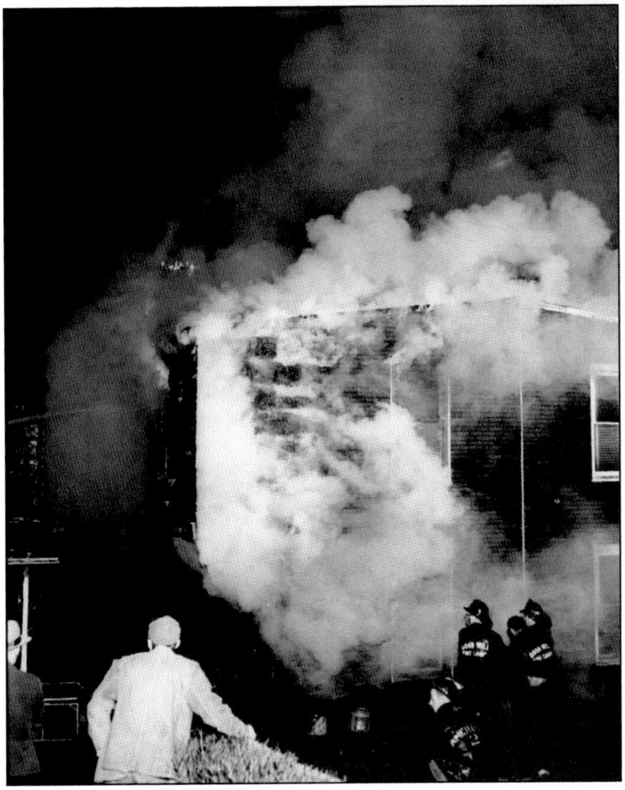

Firefighters from the American Hose of Pottsville and the Good Will Fire Company of Port Carbon operate hose lines in the rear of the homes in Palo Alto on March 14, 1957. The heavy smoke pushing from the exterior wall and window frame of the home indicates fire traveling inside the frame walls. (SHFS.)

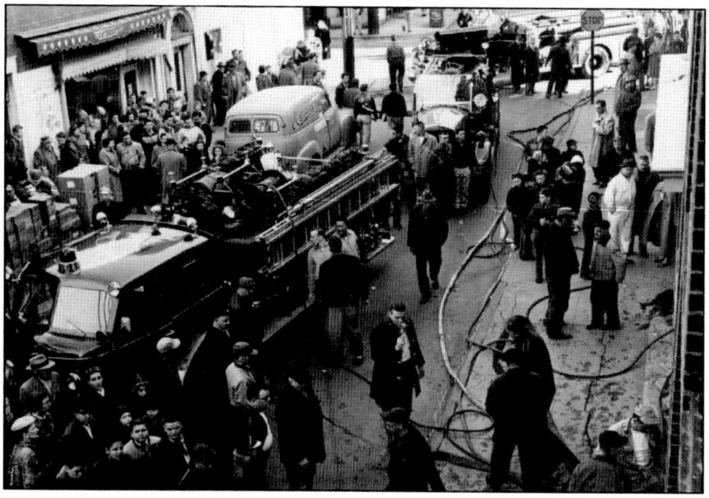

Taken from the second floor of the Good Intent Fire Company, this shot shows apparatus operating at a fire that erupted at the Trudy Shop at 5 North Second Street in Pottsville on February 21, 1957, at 3:05 p.m. Apparatus in the photograph include American Hose's 1952 American LaFrance pumper and Humane's 1929 Buffalo hose and booster truck and 1953 Seagrave pumper. (HSSC.)

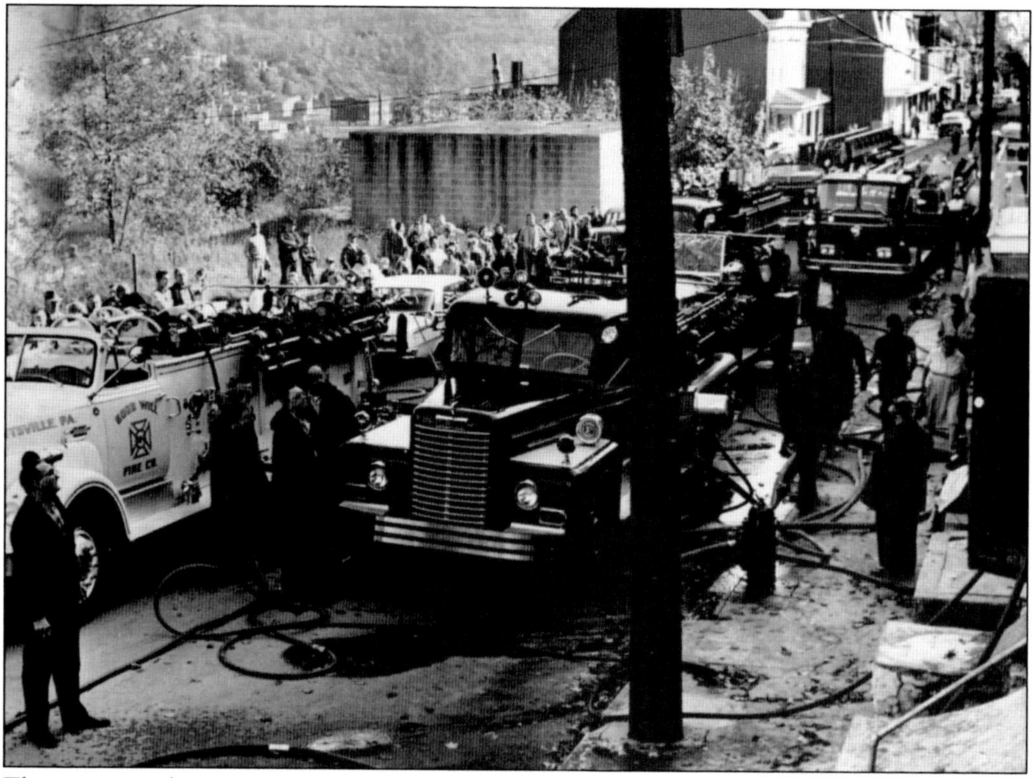

This is a great photograph of some vintage Pottsville fire apparatus. On October 27, 1962, at 2:15 p.m., fire struck the row homes at 727–729–731 North Second Street as the result of an electrical short circuit. Apparatus from six of the eight Pottsville fire companies can be readily identified in the photograph including Good Will's 1953 GMC/American 750-gallon pumper (left) and Good Intent's 1958 Oren 750-gallon pumper (right). (SHFS.)

Yorkville firefighters advance a booster line over a ground ladder to the top floor of the Eagles Country Club in the Bull's Head section of Norwegian Township. Smoke is showing from the dormer window. The alarm of fire was turned in at 1:59 a.m. on Sunday, June 9, 1963. (SHFS.)

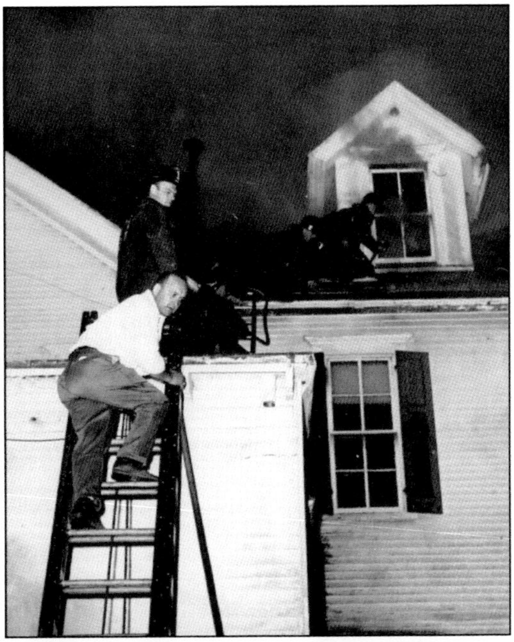

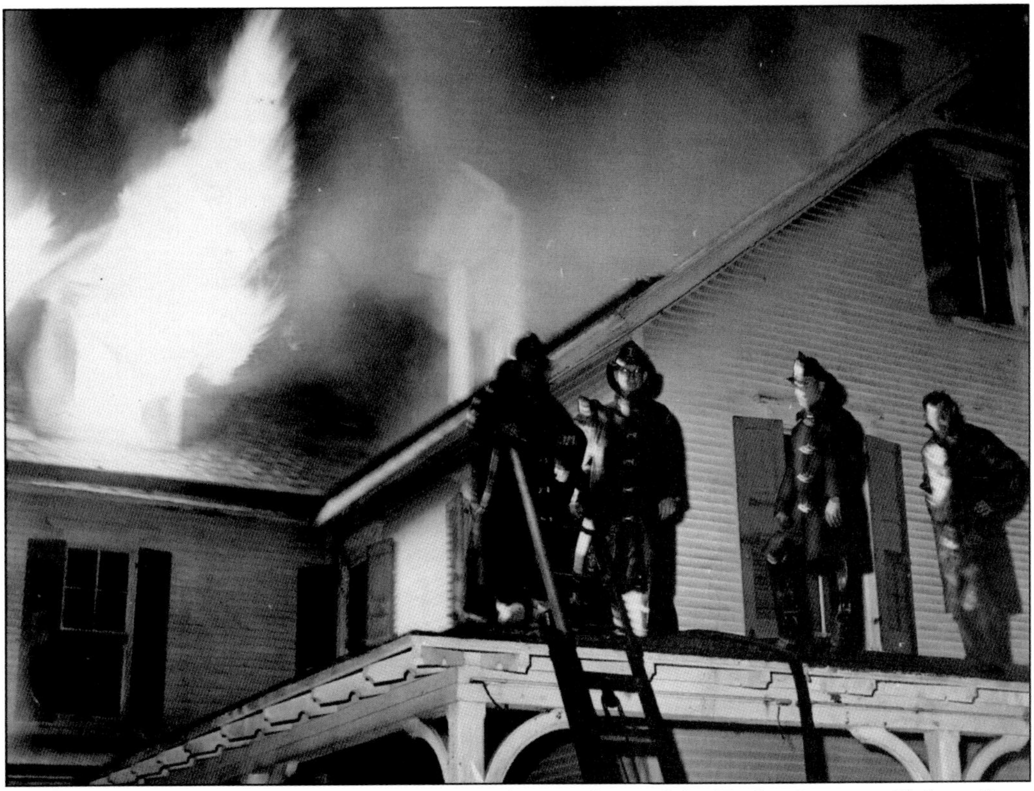

Firefighters wait for water as flames consume the top floor of the Eagles Country Club on June 9, 1963. A fire originating in the kitchen caused approximately $20,000 damage to the building. The American Hose, Humane, Yorkville, and West End fire companies of Pottsville assisted fire chief Harold Keating and the Seltzer Fire Company. (SHFS.)

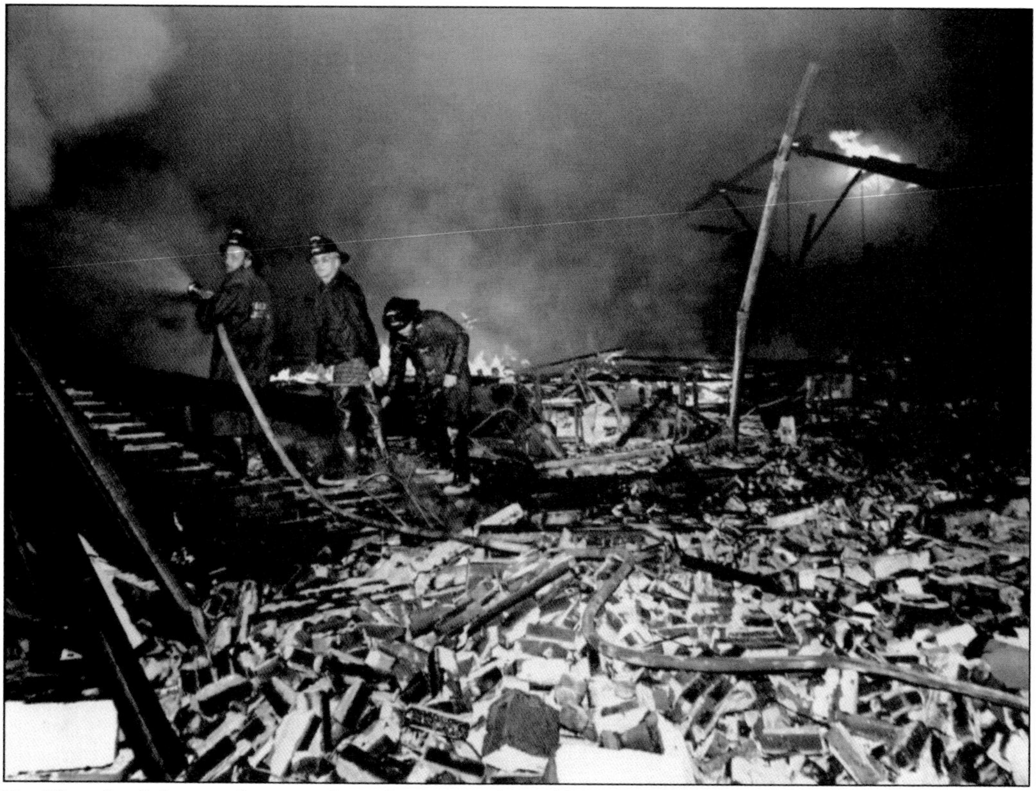

St. Clair firefighters advance a hose line over the rubble of the Clair Fashions garment factory at Mill and Russell Streets in the borough. The fire occurred on March 29, 1975. An explosion was reported to have occurred, which may account for the extensive damage and debris field. The building consisted of both a brick portion and a frame portion. (HSSC.)

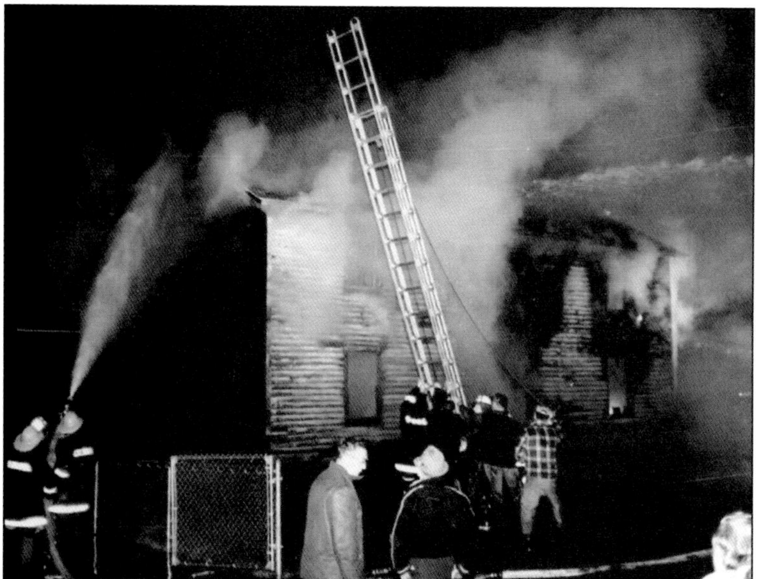

A ground ladder is raised on the Russell Street side of the Clair Fashions factory in St. Clair. Other firefighters direct a stream from a 2.5-inch hose line at the wooden portion of the building. The fire was discovered at approximately 3:00 a.m. One firefighter was injured. (HSSC.)

The abandoned former Pennsylvania Railroad roundhouse at Mount Carbon is fully involved in fire on May 1, 1961. The building had been operated as a scrap yard for some time after it was vacated by the railroad. It was located along Route 122 (currently Route 61). Pottsville firefighters pumped water from the Schuylkill River to battle the blaze. (HSSC.)

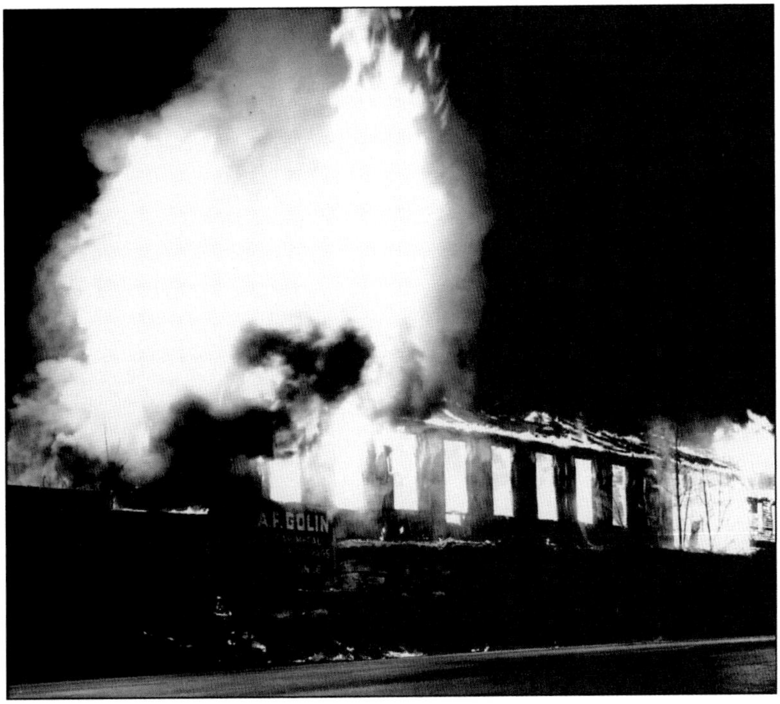

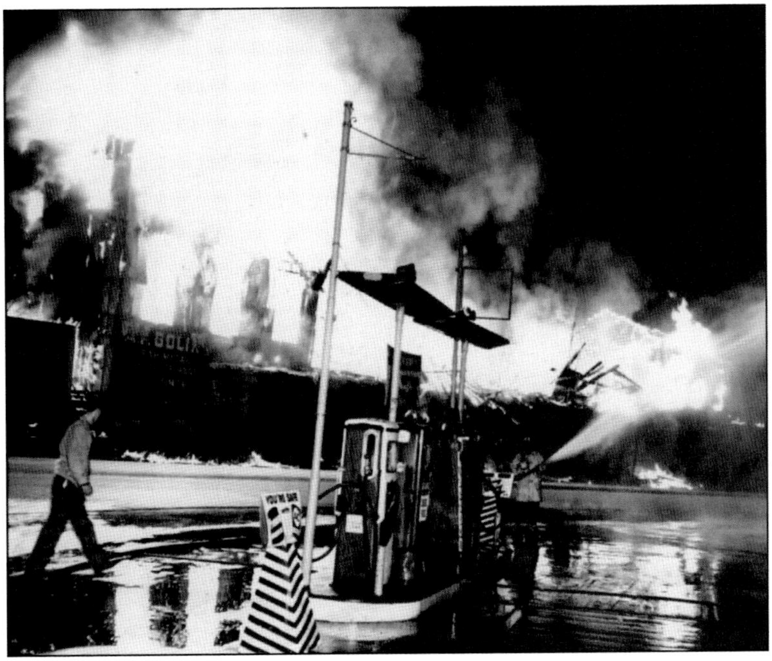

Foreman Jerry Brennan of the Good Intent Fire Company, Pottsville, directs a stream onto the Semon Service Station to protect it from the intense radiant heat from the blazing former Pennsylvania Railroad roundhouse. Four Pottsville fire companies responded to the 8:42 p.m. alarm. (HSSC.)

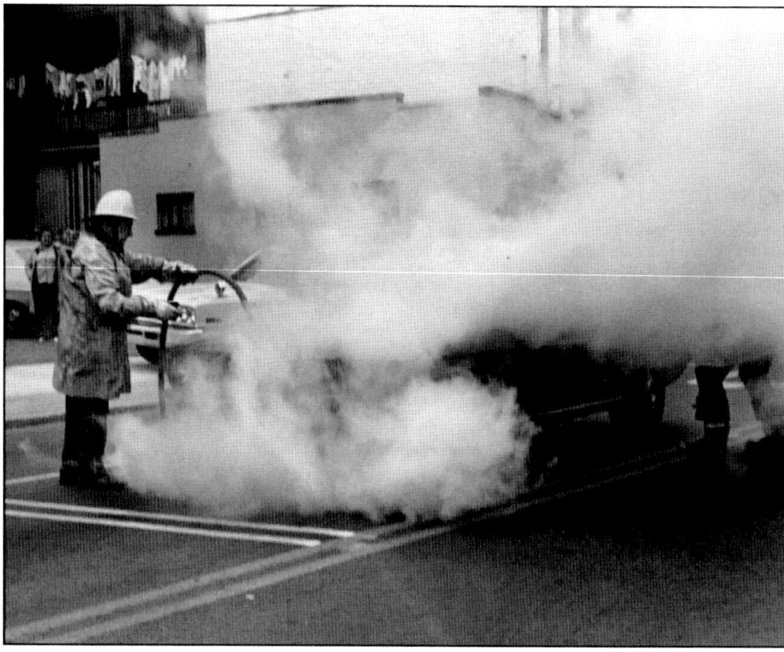

Is that a white helmet in that smoke? Pottsville fire chief Andy Hoke takes the nozzle of the booster line at this fire involving the engine compartment of a car at South Centre and Mauch Chunk Streets in downtown Pottsville on October 31, 1975. (HSSC.)

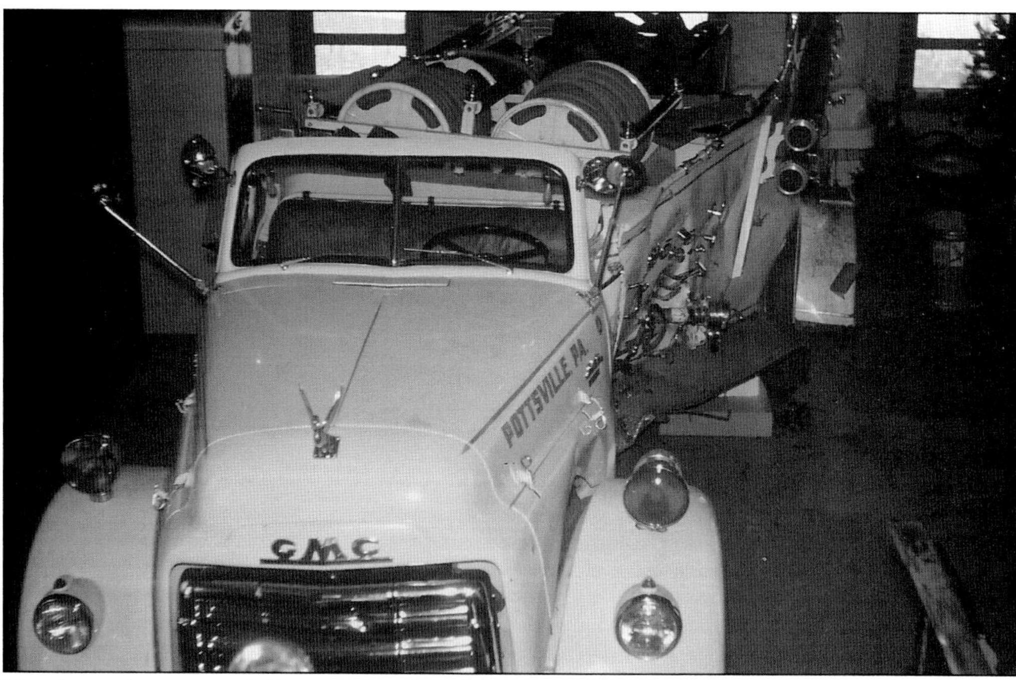

No, this isn't an optical illusion! Good Will Fire Company of Pottsville's 1953 GMC/American pumper is bent awkwardly after having been struck broadside by a coal truck at Route 122 (now Route 61) and Mill Creek Avenue in late 1962. Several firefighters were injured but none seriously. The truck was beyond repair. (William Reigle; photograph by Jack Yaag.)

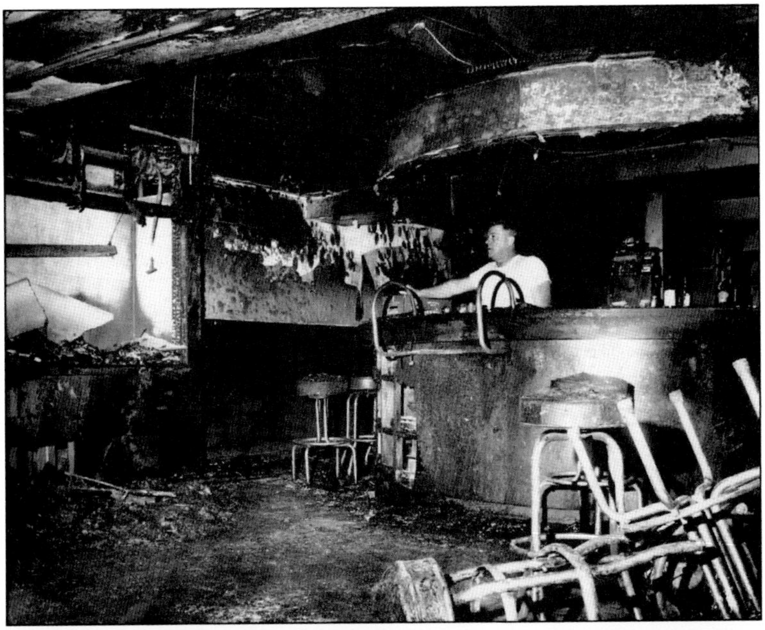

The steward of the American Legion post in Port Carbon surveys the damage to the bar area of the building following a fire on June 20, 1964, at 4:50 a.m. Two firefighters from Pottsville were injured when a smoke explosion hurled one of them through a third-floor window and he struck a colleague coming to his aid. (HSSC.)

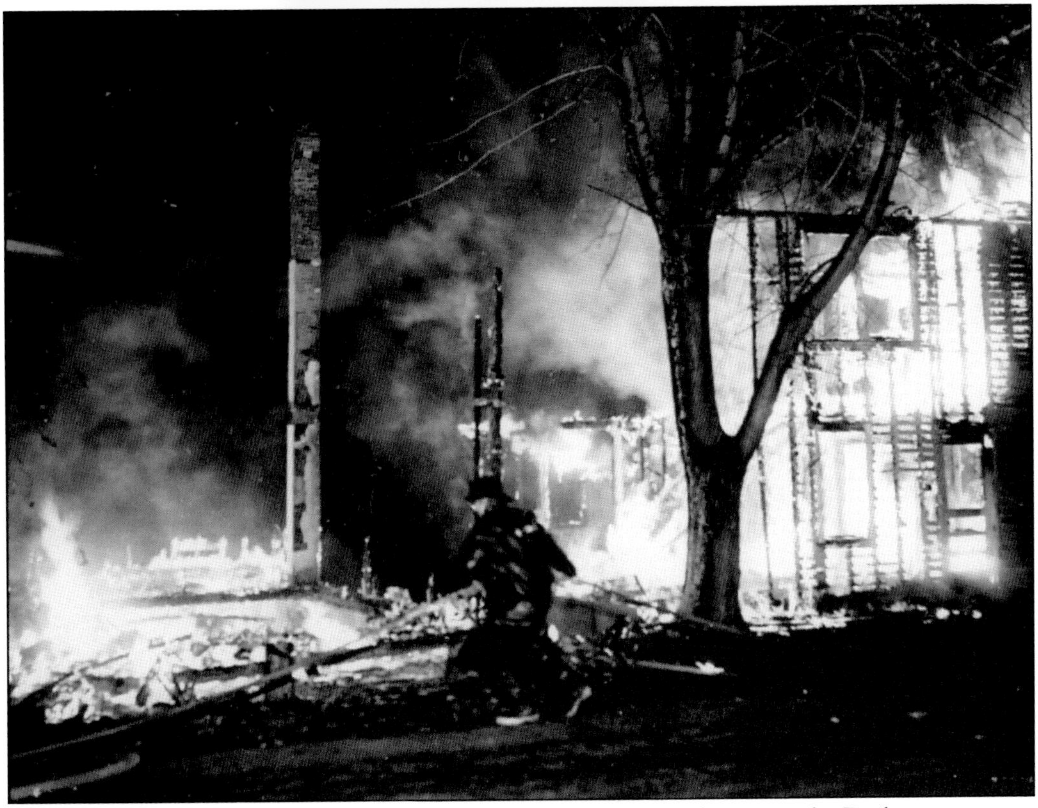

A firefighter advances a hose line as flames consume these frame homes in the Darkwater section of New Castle Township. Three two-story frame homes were leveled and one other was damaged on November 11, 1965. Water was relayed by fire department pumpers over 1 mile to the scene. Pottsville and St. Clair firefighters responded to the 4:30 a.m. alarm. (HSSC.)

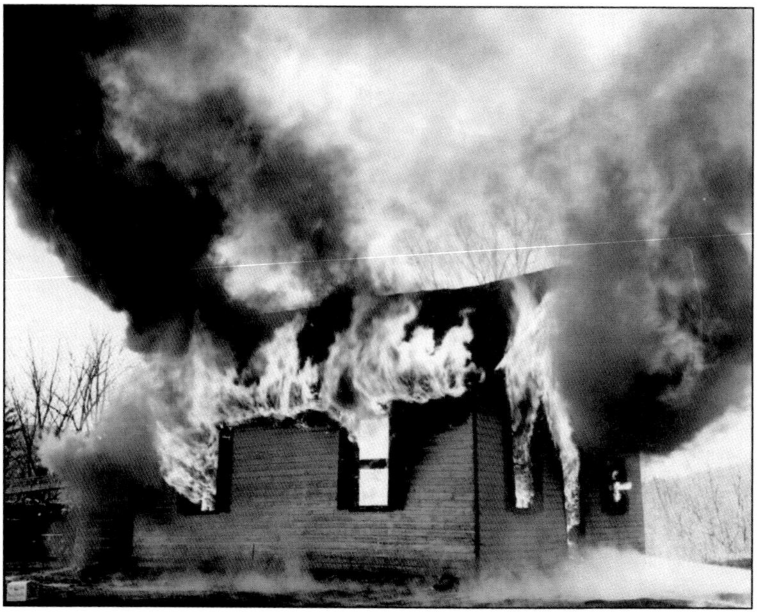

Flames engulf the 100-year-old Seiders Beer Hall tavern on the Gordon Nagle Trail on the outskirts of Pottsville on April 11, 1974. The fire was discovered at approximately 8:00 a.m. Barely visible to the left in the photograph is West End of Pottsville's 1960 Seagrave 65-foot quint. Firefighters from Cressona and Pottsville fought the fire. (HSSC.)

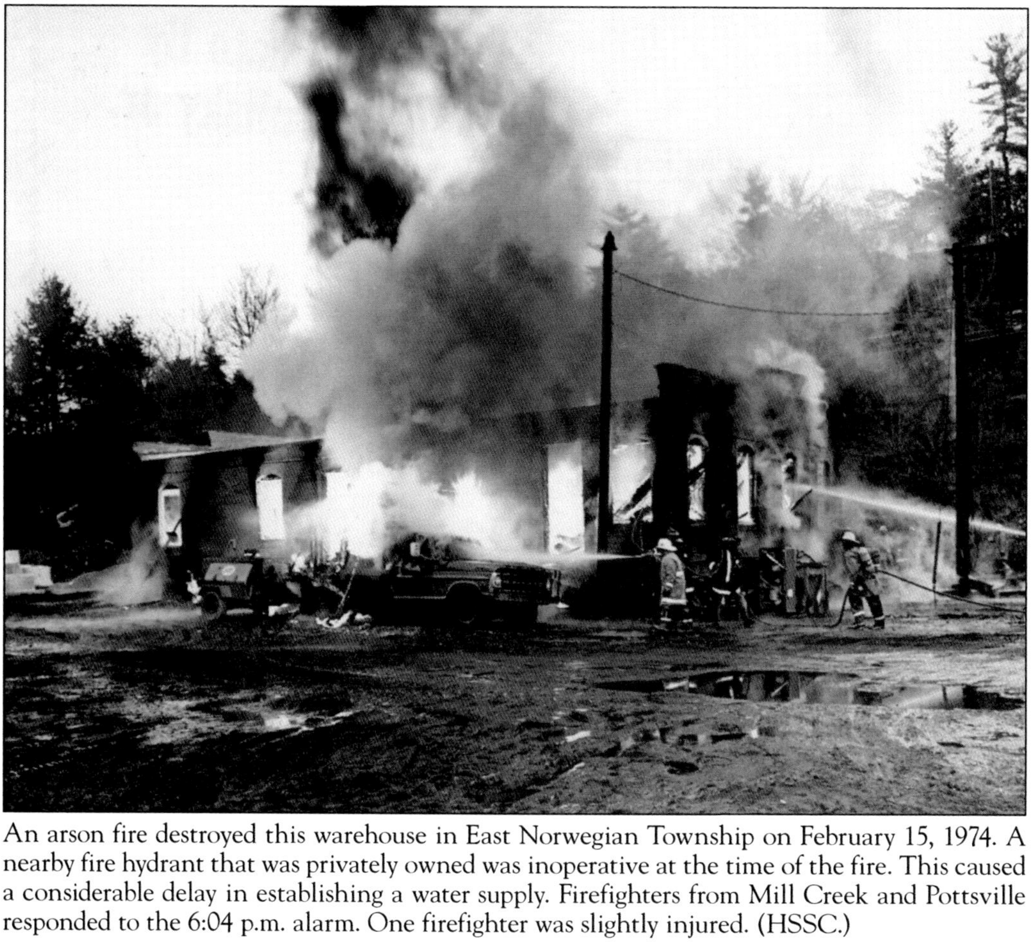

An arson fire destroyed this warehouse in East Norwegian Township on February 15, 1974. A nearby fire hydrant that was privately owned was inoperative at the time of the fire. This caused a considerable delay in establishing a water supply. Firefighters from Mill Creek and Pottsville responded to the 6:04 p.m. alarm. One firefighter was slightly injured. (HSSC.)

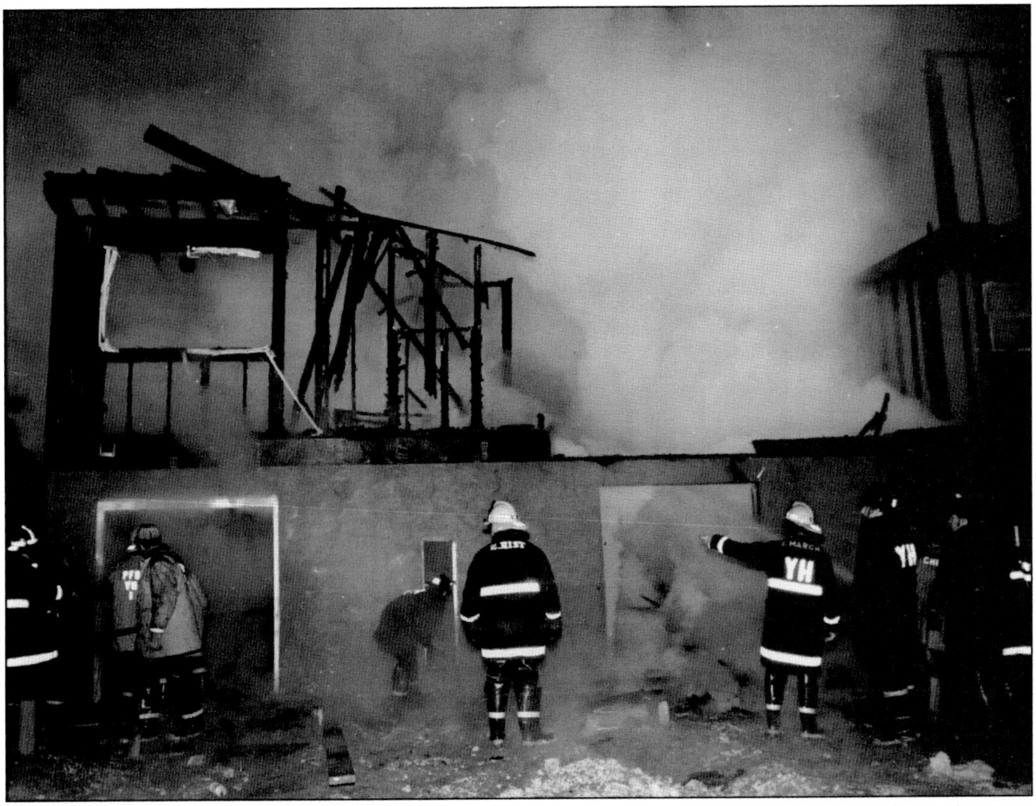

On October 21, 1974, fire destroyed three newly constructed garden apartments along the Mar Lin Road in Norwegian Township, just outside Pottsville. Pottsville firefighters responded to the alarm at 2:40 a.m. Damage was estimated at $60,000. The firefighter from Yorkville Hose (YH) with "T. March" on the back of his turnout coat was destined to become fire chief in Pottsville a little over a year later and continues to serve today. (HSSC.)

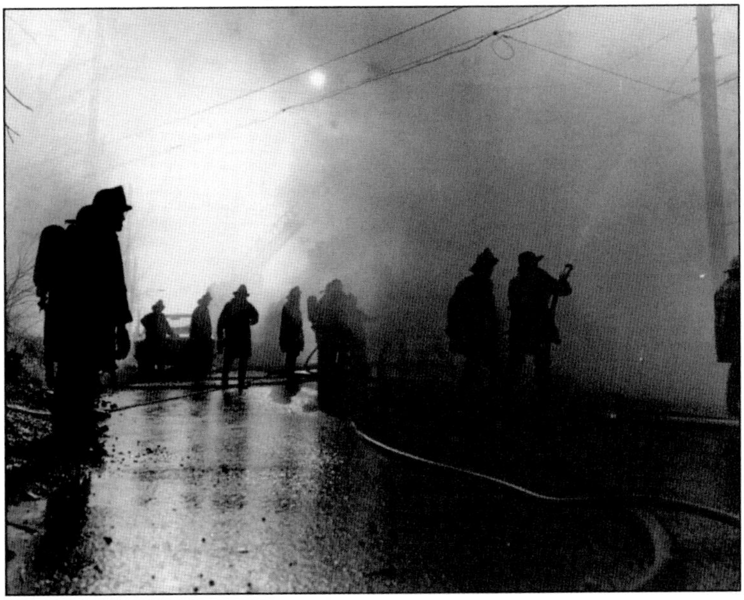

Firefighters are silhouetted in the smoke at this fire that gutted three row homes at 452–454–456 Adams Street in Pottsville on March 7, 1974. Box 15 was transmitted at 4:24 p.m. Firefighters arrived to find heavy fire showing from the rear of the three houses. A home at 15 South Jackson Street was damaged by radiant heat. The fire caused $39,750 in damage. (HSSC.)

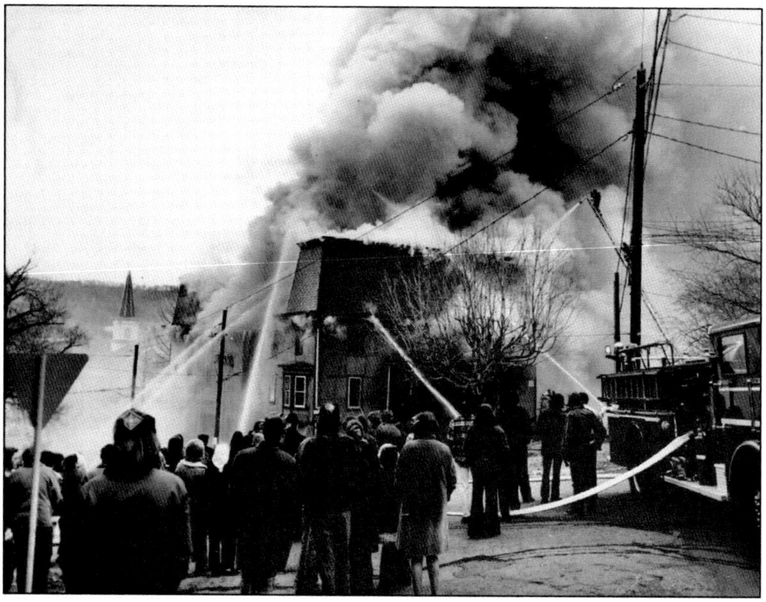

On February 24, 1974, flames destroyed the Angela Sportswear factory in Port Carbon. The alarm was sounded at approximately 5:15 p.m. Good Will of Port Carbon operates a 1968 Ward LaFrance pumper as numerous hose-line streams are directed at the building. West End of Pottsville has a ladder pipe in operation to the right of the building. (HSSC.)

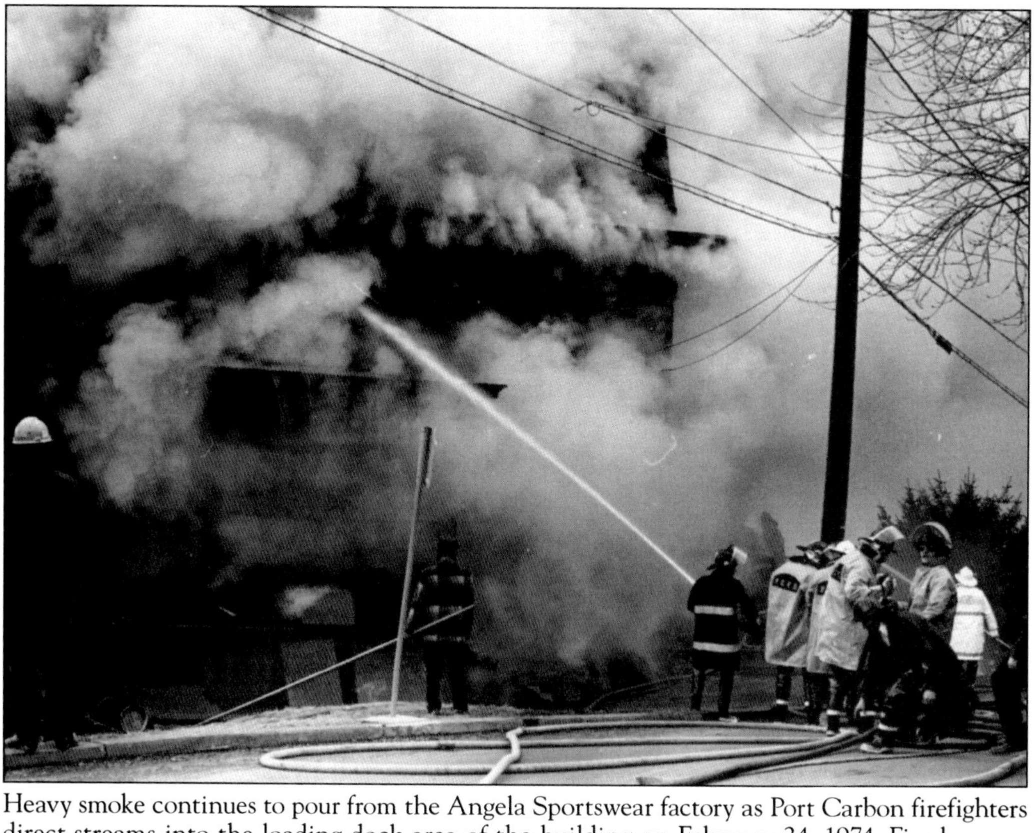

Heavy smoke continues to pour from the Angela Sportswear factory as Port Carbon firefighters direct streams into the loading dock area of the building on February 24, 1974. Fire loss was estimated at $150,000. Eighty individuals lost their jobs as a result of the fire. The factory was located in the former Ritz Theater at First and Washington Streets. (HSSC.)

Good Will Fire Company of Pottsville has a 1963 GMC/American pumper in service at the Angela Sportswear factory fire in Port Carbon on February 24, 1974. Heavy smoke pours from the upper floors of the factory as flames begin to break through the roof. (Good Will Fire Company, Pottsville; photograph by Robert Dusel.)

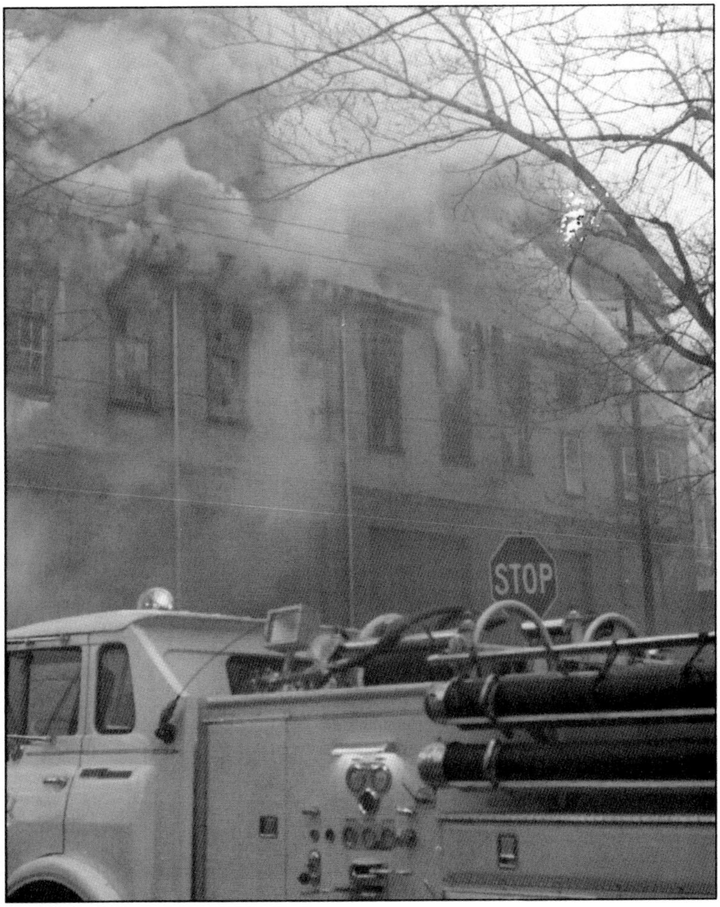

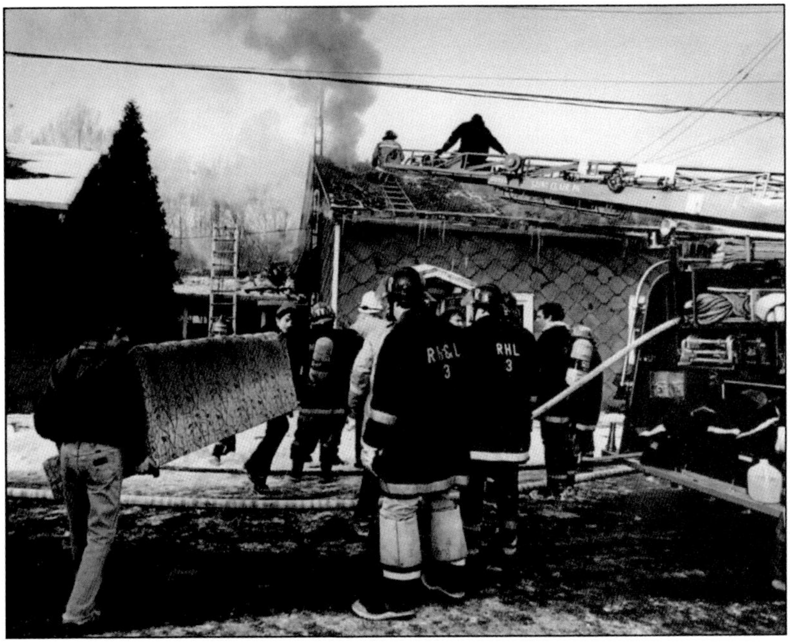

Firefighters open the roof from the 65-foot aerial ladder of the Rescue Hook and Ladder Company of St. Clair at this fire on January 20, 1982. The alarm was transmitted to St. Clair firefighters at 8:15 a.m. for the two-story frame home at 396 North Mill Street. The fire began in a summer kitchen. In the photograph, a sofa salvaged from the home is removed. (HSSC.)

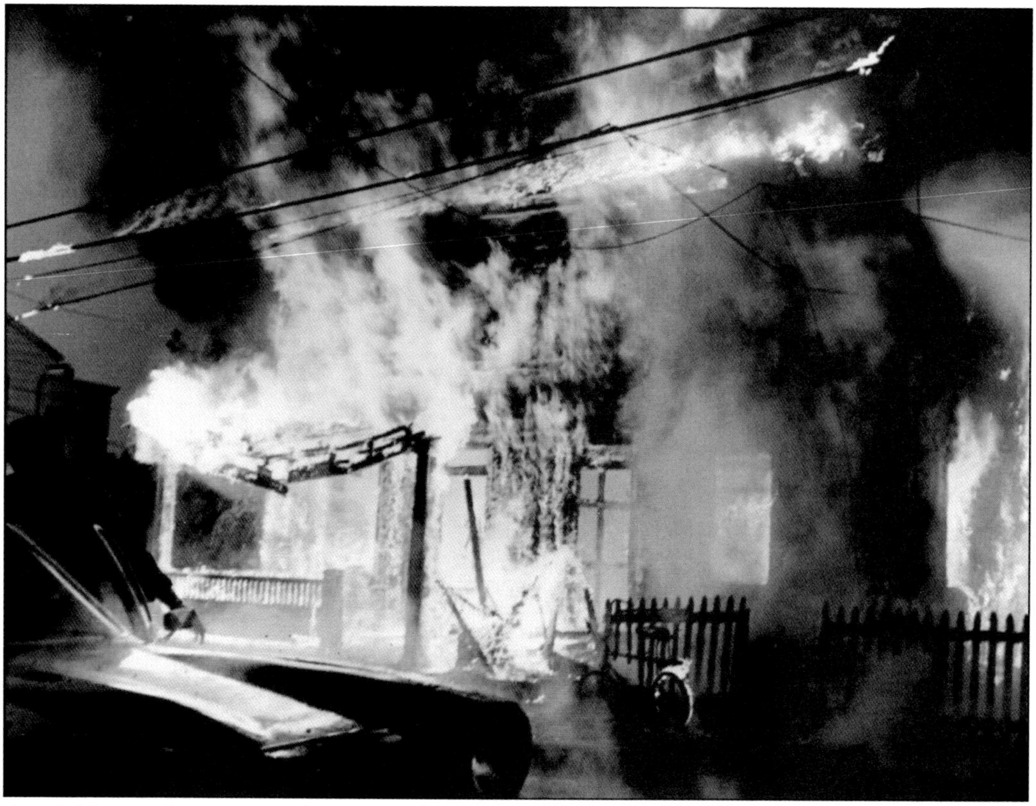

A neighbor rushes to move his car as flames engulf this double-block frame home at 259–261 Savory Street in Palo Alto on May 11, 1978. Overhead utility wires are burning, as is a bicycle in front of the fence. A 22-year-old male who resided at 259 Savory Street was killed in the fire. (HSSC.)

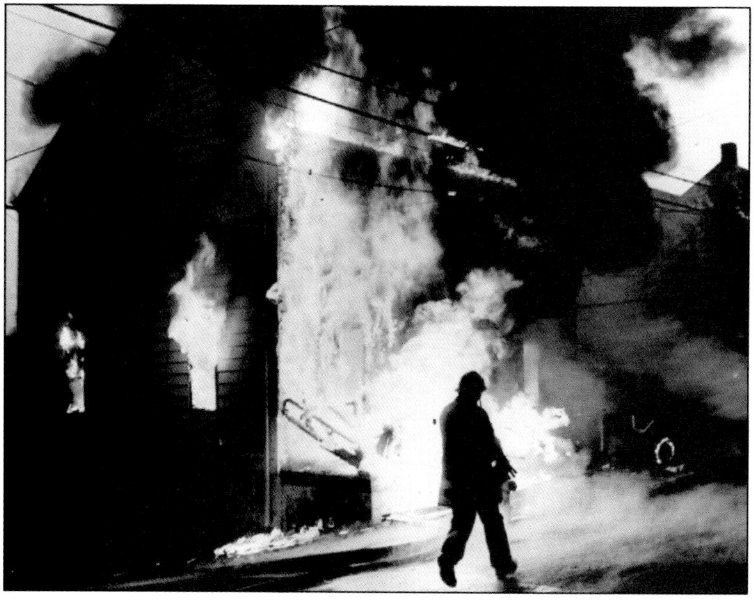

The front porches of the homes on Savory Street in Palo Alto have collapsed as firefighters arrived on the scene. A night watchman from a garment factory in downtown Pottsville noticed the smoke at about 5:30 a.m. He notified Pottsville Police. Upon further investigation, police discovered the fire in Palo Alto and transmitted the alarm. (HSSC.)

A hose line snakes in the street to the rear of American Hose of Pottsville's 1975 American LaFrance pumper, which is operating at the fire on Savory Street in Palo Alto. Firefighters from Port Carbon and Pottsville assisted Palo Alto firefighters on May 11, 1978. (HSSC.)

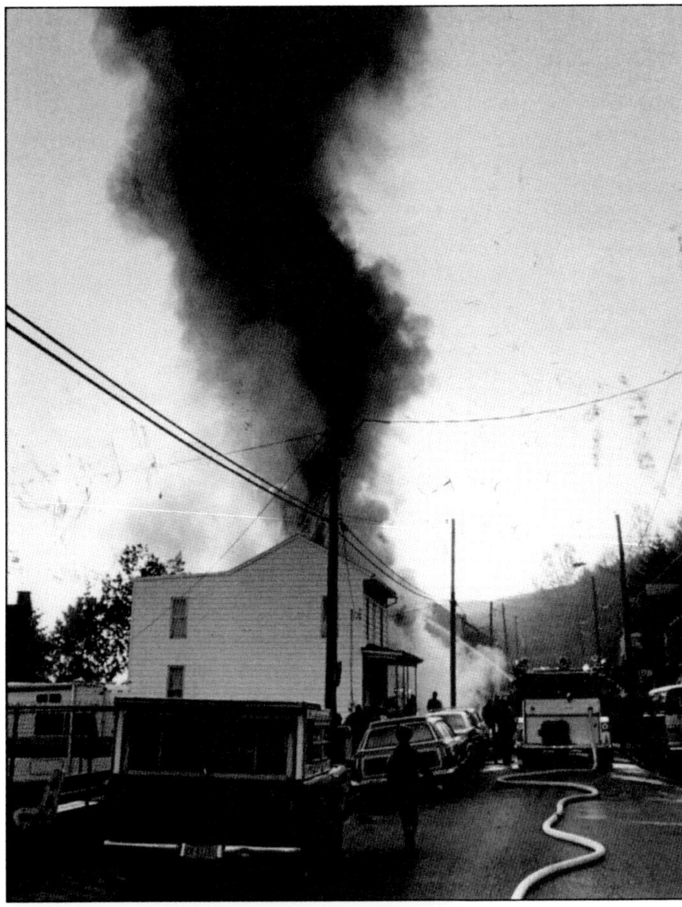

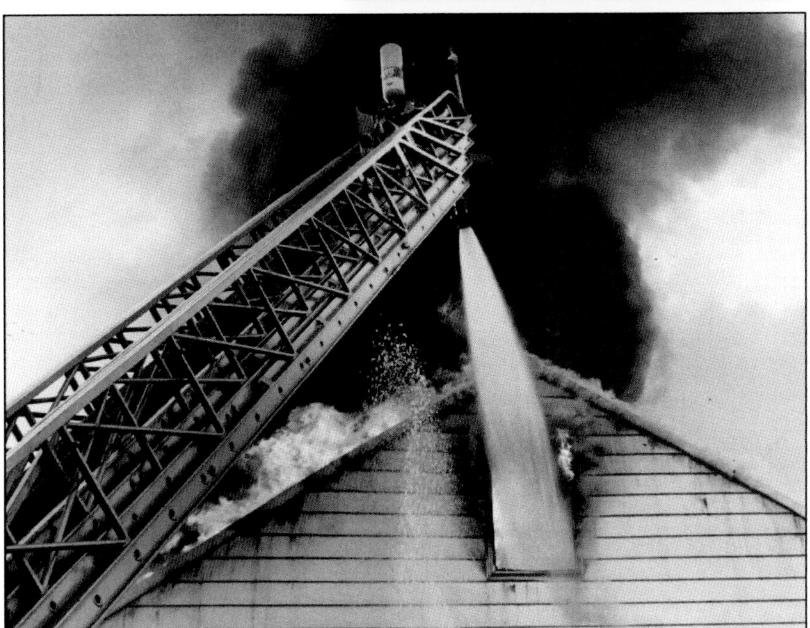

A firefighter directs a stream from the ladder pipe into the attic of the fully involved frame duplex on Savory Street. West End of Pottsville's 1977 Seagrave 100-foot quint was relatively new when it was placed in service at the fire in Palo Alto in May 1978. (HSSC.)

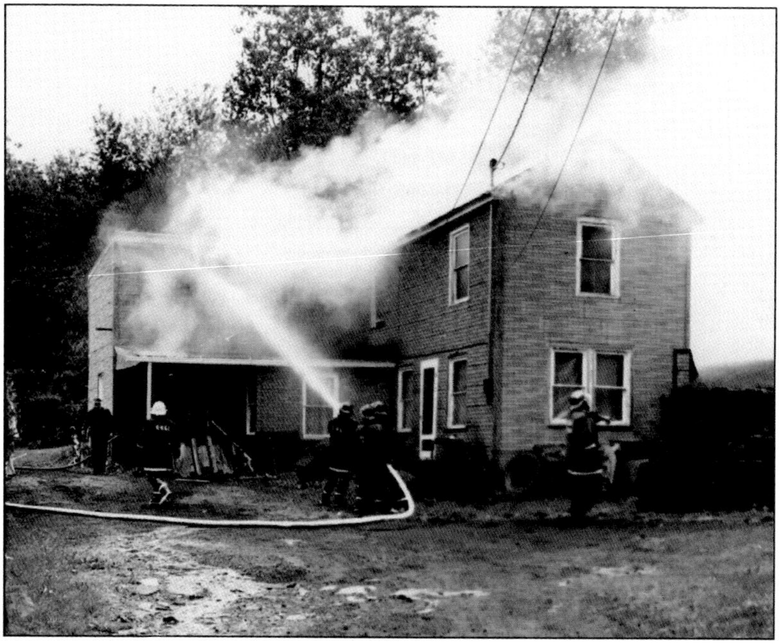

Fire originating in the second-floor rear bedroom of this frame home caused extensive damage to the second floor and attic on June 1, 1976. The blaze in Wadesville was discovered at approximately 10:00 a.m. Firefighters from Wadesville and St. Clair fought the fire. (HSSC.)

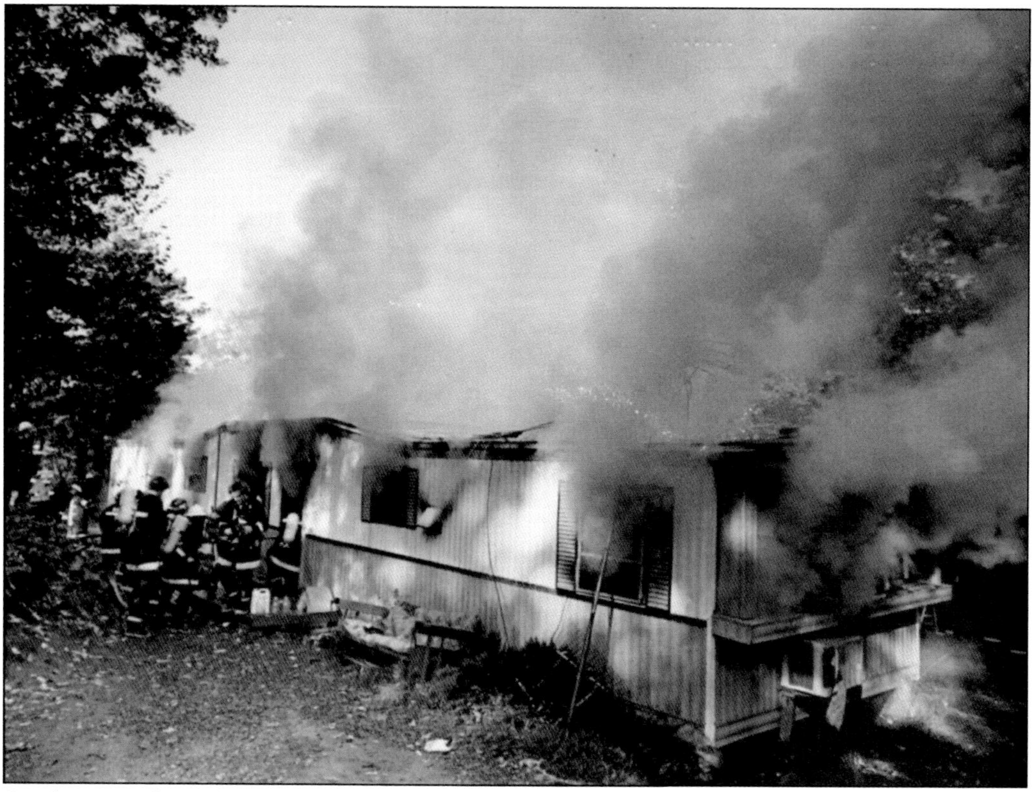

Smoke pours from this trailer home in Seider's Trailer Park in North Manheim Township, just outside of Pottsville. The fire occurred on October 14, 1983, at 9:15 a.m. Firefighters push into the front door to knock down the flames. Fire companies from Pottsville, Cressona, and Schuylkill Haven fought the fire. (HSSC.)

Hose lines are stretched from Alert Fire Company of St. Clair's 1968 Mack pumper at this fire involving a large number of illegally dumped tires in an abandoned strip-mining pit in Blythe Township, just outside of St. Clair. The alarm was transmitted at 9:30 a.m. on May 13, 1983. A large water tanker owned by the Reading Anthracite Coal Company was used to supply water to fight the blaze. (HSSC.)

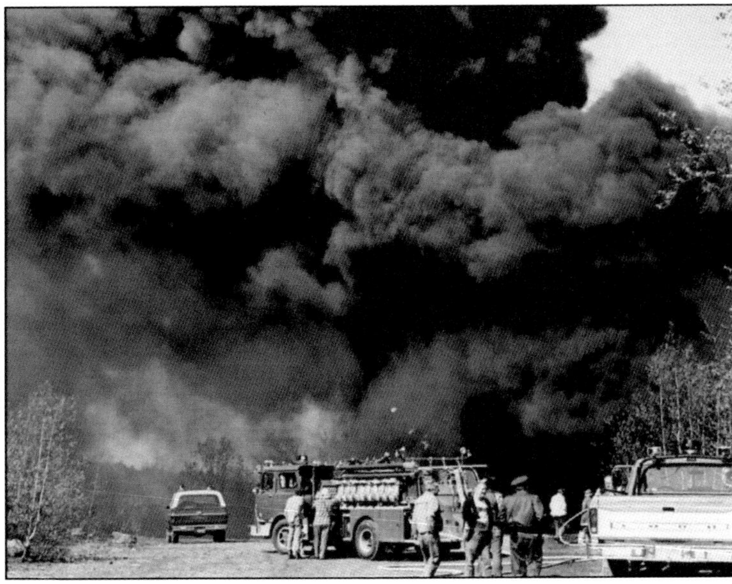

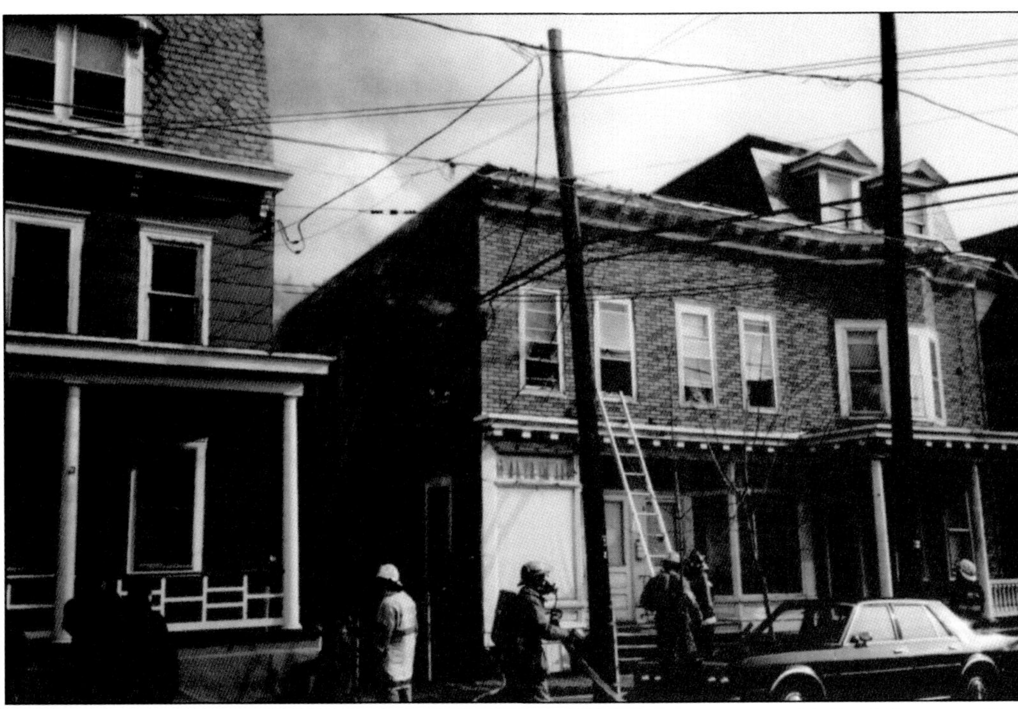

On November 15, 1992, fire broke out in this second-floor apartment at 66 Coal Street in Port Carbon. Firefighters stretch hose lines as fire begins to vent from the windows of the frame building. Members of Port Carbon, Palo Alto, and the Phoenix and Yorkville companies of Pottsville contained the flames to the apartment of origin. (Authors' collection.)

SCHUYLKILL COUNTY FIRE COMPANIES BY MUNICIPALITY

CITY	DATE	STATUS
Pottsville		
Pottsville Fire Company No. 1	1829	Renamed Humane Hose Company No. 1, Active
Schuylkill Hydraulians	1829	Renamed Phoenix Fire Engine Company No. 2, Active
Good Intent Fire Company No. 1	1846	Active
Rough and Ready Fire Company	1848	Renamed American Hose Company No. 2, Active
Atkins Fire Company	1873	Disbanded—1894
Good Will Fire Company No. 4	1892	Active
Greenwood Hill Fire Company	1946	Disbanded—2008
West End Fire Company No. 7	1886	Active
Yorkville Hose and Fire Company No. 1	1891	Active
Pottsville Junior Fire Companies—mid-19th century (*All Pottsville boy (junior) fire companies were banned after the Civil War.*)		
Rangers Boy Fire Company		Sponsored by the Rough and Ready Fire Company (American Hose Company)
Western Boy Fire Company		Sponsored by the Schuylkill Hydraulians
Young America Fire Company		Sponsored by the Good Intent Fire Company

BOROUGHS	DATE	STATUS
Ashland		
American Hose Company	1878	Active
Washington Fire Company	1867	Active
Auburn		
Vigilant Fire Company No. 1	1898	Reorganized 1921
Auburn Fire Company No. 1	1921	Active
Coaldale		
Coaldale Volunteer Fire Company No. 1	1906	Active
Cressona		
Cressona Fire Company No. 1	1899	Active
Good Will Fire Company	1910	Active
Deer Lake		
Deer Lake Fire Company	1944	Renamed Deer Lake and West Brunswick Township Fire Company, 1956, Active
Frackville		
Good Will Hose Company No. 1	1910	Active
Gilberton Borough		
Friendship Hose Company No. 1, Maizeville	1886	Disbanded in the 1930s
American Hose Company No. 2, Mahanoy Plane	1886	Active
Continental Hose Company No. 3, Gilberton	1886	Active
Girardville		
Girard Hose Company No. 1	1882	Active
Rangers Hose Company	1911	Active
West End Hose Company	1917	Disbanded 1945
Gordon		
Citizens Fire Company	1927	Active
Landingville		
Community Fire Company, Landingville	1950	Active
Mahanoy City		
Humane Fire Company No. 1	1867	Active
Citizens Fire Company No. 2	1870	Active
Washington Hook and Ladder Company No. 1	1886	Active
Good American Hose Company No. 3	1893	Active
West End Fire Company No. 5	1925	Active
McAdoo		
Keystone Fire Company	1896	Active
McAdoo Fire Company	1921	Active
Mechanicsville		
Mechanicsville Hose Company	1903	Active
Middleport		
Citizens Fire Company	1903	Active
Minersville		
Mountaineer Hose Company	1861	Active
Good Will Fire Company	1867	Active
Rescue Hook and Ladder Company	1871	Active
Independent Hose Company	1869	Active
New Minersville Fire Company	1926	Active
Minersville Fire and Rescue	2009	Merger of Rescue and Independent companies, Active
Mount Carbon		
Mount Carbon Fire Company No. 1	1910	Renamed North Manheim Township Volunteer Fire Company No. 1, 2007, Active
New Philadelphia		
Good Intent Fire Company	1905	Active
New Ringgold		
Community Fire Company	1926	Active
Orwigsburg		
Orwigsburg Fire Company	1813	Chartered in 1887 as the Friendship Hose Company No. 1
Friendship Hose Company No. 1	1887	Active
Palo Alto		
Citizens Fire Company	1884	Active
East End Fire Company	1907	Active
Pine Grove		
Pine Grove Hose Company	1855	Renamed to Hose, Hook, and Ladder Company No. 1, Active
North End Fire Company	1935	Active
Port Carbon		
Good Will Fire Company No. 1	1879	Active
Port Clinton		
Port Clinton Fire Company	1946	Active
Ringtown		
Chemical Fire Company	1898	Merged in 2007 to become Ringtown Valley Fire and Rescue
Ringtown Valley Fire and Rescue	2007	Active
St. Clair		
Alert Fire Company No. 1	1883	Active
Columbia Hose Company No. 2	1868	Active
Phoenix Fire Company, East Mines	1911	Active
Rescue Hook and Ladder Company	1898	Active
West End Fire Company, Arnots Addition	1951	Active
Schuylkill Haven		
Good Intent Fire Company No. 1	1885	Renamed Spring Garden Hose Company, then Rainbow Hose Company No. 1, Active
Schuylkill Hose Company No. 2	1892	Active
Friendship Hook and Ladder Company No. 3		Merged in 1910 with Rainbow Hose Company
Liberty Fire Company No. 4	1908	Active
Shenandoah		
Columbia Hose and Steam Fire Engine Company No. 1	1870	Active
Rescue Hook and Ladder Company No. 1	1873	Active

Phoenix Fire Company No. 2	1885	Active
Defender Hose Company No. 3	1895	Active
Polish-American Fire Company No. 4	1915	Active
Tamaqua		
Reliance Hook and Ladder Company	1860	Renamed American Hose Company No. 1, 1878, Active
Perseverance Hose and Steam Fire Engine Company	1879	Renamed Citizens Fire Company, 1889, Active
East End Fire Company	1916	Active
South Ward Fire Company	1913	Active
Tower City		
Tower City Fire Company No. 1	1899	Active
Tremont		
Tremont Fire Company No. 1	1878	Active

TOWNSHIPS	DATE	STATUS
Branch		
Good Intent Fire Company, Llewellyn	1908	Active
Phoenix Park Fire Company	1966	Active
Blythe		
Good Will Hose Company, Cumbola	1899	Active
Kaska Fire Company	1957	Active
Butler		
Friendship Fire Company, Englewood	1925	Active
American Fire Company, Fountain Springs	1924	Active
Lavelle Volunteer Fire Company	1937	Active
Locust Dale	1935	Active
Cass		
Clover Hose Company, Heckscherville	1938	Active
Duncott Hose Company No. 1	1951	Active
East Norwegian		
American Fire Company, Schoentown	1927	Active
Independent Hose Company	1915	Active
Mill Creek Hose Company		Renamed East Norwegian Township Emergency Services, Active
East Union		
Sheppton-Oneida Fire Company No. 1	1933	Active
Citizens Fire Company, Brandonville	1936	Active
Eldred		
Sacramento Fire Company	1959	Active
Foster		
Mackeysburg Fire Company	1936	Disbanded
Mount Pleasant Hose Company No. 1, Buck Run	1947	Active
Frailey		
Donaldson Fire Company	1926	Active
Hegins		
Hegins Fire Company	1923	Active
Valley View Fire Company	1927	Active
Kline		
Kline Township Volunteer Fire Company	1946	Disbanded in the 1990s
Mahantongo		
Mahantongo Fire Company, Pitman	1951	Active
New Castle		
Rainbow Hose Company, Wadesville	1920	Active
North Manheim		
North Manheim Volunteer Fire Company No. 1	2007	Active
North Union		
Nuremberg/Weston Volunteer Fire Company	1934	Active
Norwegian		
Mar Lin Citizens Hose Company	1932	Active
Seltzer Hose Company	1945	Active
Pine Grove		
Suedberg Volunteer Fire Company	1948	Active
Ravine	1946	Active
Porter		
Joliett Fire Company	1956	Active
Reiner City Hook and Ladder	1922	Renamed Muir Fire Company, 1948, Active
Orwin Fire Company	1920	Active
Reinerton Fire Company	1926	Active
West End Fire Company No. 3, Sheridan	1926	Active
Reilly		
Citizens Fire Company, Branch Dale	1920	Active
Newtown Volunteer Fire Company, Zerbe	1954	Active
Ryan Township		
Ryan Township Fire Company	1940	Active
Rush Township		
Hometown Fire Company	1941	Active
Quakake Volunteer Fire Company	1941	Active
Schuylkill		
Brockton Fire Company	1913	Active
Mary-D Fire Company	1950	Active
Tuscarora Fire Company	1923	Active
South Cass		
Citizens Fire Company	1945	Active
Union		
Union Township Fire Company No. 1	1936	Merged in 2008 to become Ringtown
Valley Fire Company		
Upper Mahantongo		
Klingerstown Fire Company	1949	Active
Walker		
New England Fire Company	1953	Active
Wayne		
Friedensburg Fire Company	1960	Active
Summit Station Fire Company	1948	Active
West Mahanoy		
Altamont Fire Company	1931	Active
Citizens Fire Company, Lost Creek	1922	Inactive
Heights Fire Company, Shenandoah Heights	1923	Active
Lost Creek West End Fire Company	1919	Disbanded
William Penn Fire Company, Shaft	1907	Active
West Penn		
Andreas Fire Company No. 1	1942	Active
West Penn Fire Company	1998	Active